EQUIVALENCE CHECKING OF DIGITAL CIRCUITS

Equivalence Checking of Digital Circuits

Fundamentals, Principles, Methods

by

Paul Molitor
Martin-Luther-University Halle-Wittenberg, Germany

and

Janett Mohnke
DResearch Digital Media Systems GmbH, Germany

Guest Authors

Bernd Becker, University Freiburg, Germany

Christoph Scholl, University Heidelberg, Germany

KLUWER ACADEMIC PUBLISHERS
BOSTON / DORDRECHT / LONDON

A C.I.P. Catalogue record for this book is available from the Library of Congress.

ISBN 1-4020-7725-4

Published by Kluwer Academic Publishers,
P.O. Box 17, 3300 AA Dordrecht, The Netherlands.

Sold and distributed in North, Central and South America
by Kluwer Academic Publishers,
101 Philip Drive, Norwell, MA 02061, U.S.A.

In all other countries, sold and distributed
by Kluwer Academic Publishers,
P.O. Box 322, 3300 AH Dordrecht, The Netherlands.

Printed on acid-free paper

Printed in the Netherlands.

To our families

Ursula, Levente, and Louise

Klaus, Jasmin, and Jonas

Contents

Acknowledgments

Our special thanks go to Christoph Scholl from University of Freiburg for his terrific work on reading and improving the draft of this monograph. He did a great job.

We also would like to thank Riccardo Forth who helped us with his comments.

Since this book has been planned as a textbook, we have included research results from literature that we felt were particularly relevant. We regret that we may have overlooked some important information and we would gratefully like to acknowledge the work of all the researchers who investigated time and effort to significantly develop the wide field of equivalence checking of digital circuits. Without you we would not have been able to assemble this book.

Last but not least, we would like to thank our spouses, Ursula and Klaus, for their persistent encouragement and their patience during our work on this book.

Alone we can do so little, together we can do so much — Helen Keller

Chapter 1

INTRODUCTION

Hardware verification is the process of checking whether a design conforms to its specifications of functionality and timing. In today's design processes it becomes more and more important. Very large scale integrated (VLSI) circuits and the resulting digital systems have conquered a place in almost all areas of our life, even in security sensitive applications. Complex digital systems control airplanes, have been used in banks and on intensive-care units. Hence, the demand for error-free designs is more important than ever. In addition, economic reasons underline this demand as well. The design and production process of present day VLSI-circuits is highly time- and cost-intensive. Moreover, it is nearly impossible to repair integrated circuits. Thus, it is desirable to detect design errors early in the design process and not just after producing the prototype chip. All these facts are reflected by developing and production statistics of present day companies. For example, Infineon Technologies [118] assumed that about 60% to 80% of the overall design time was spent for verification in 2000. Other sources cite the 3-to-1 head count ratio between verification engineers and logic designers. This shows that verifying logical correctness of the design of hardware systems is a major gate to the problem of time-to-market (cf. [113]).

With the chip complexity constantly increasing, the difficulty as well as the importance of functional verification of new product designs has been increased. It is not only more important to get error-free designs. Moreover, it becomes an increasingly difficult task for a team of human designers to carry out a full design without errors. The traditional training of new verification engineers has to be adapted to the new situation. New skills are necessary. As mentioned before, it is necessary to find design errors early in the design process. It is important to develop skills to handle design processes which may lead into functionally incorrect designs. For these reasons, nearly all major universities offer lectures on basic verification techniques such as propositional temporal logic, model checking, equivalence checking, and simulation coverage measures. The present book is designed as a textbook covering one of the most important aspects in the verification process — *equivalence checking of Boolean circuits*.

1. Tasks in verification processes

The validation process of a design mainly consists of three tasks in today's verification processes:

At first, the designers have to give a formal specification which describes the desired behavior of the circuit. Usually, such a specification is given by a set of properties, each of them expressed in a propositional temporal logic [48]. The designers have to assure themselves of the completeness of their specification, i.e., they have to answer the question 'What properties do we really want to prove about this circuit?'. All relevant aspects of a design have to be covered by the specification. This 'Have we written enough properties?'-problem is one subject of current international research [28, 38, 71].

After having designed a first implementation of the circuit, which is usually given at the register-transfer level, it has to be proven that the specified properties are satisfied by the implementation, i.e., that the design is logical correct. This is done by applying a number of innovative tools and methodologies, for example, symbolic model checking [30, 100], bounded model checking [10], and test environments (cf. [8]).

Starting from a register-transfer-level description, a design is synthesized to a gate level netlist. This netlist is then subject to layout synthesis. After all, there is almost no design where not some manual corrections need to be done. During many years, it has been argued that synthesis is correct by construction, but experience shows that this is not always the case [118]. Today's synthesis systems are huge and complicated software constructs. We cannot assume that the correctness of such a system can ever be guaranteed or even be proven. Instead of attempting to formally verify the design automation tools, a more practical way is to formally check that a circuit synthesized by a design automation tool is functionally equivalent to the initial register-transfer-level description. This is the task of *equivalence checking*, namely proving that two given circuits are equivalent, i.e., that they have the same input/ouput behavior. From the industrial point of view, *equivalence checking* is the most important formal verification technique being employed in today's design flows.

2. Equivalence checking

Equivalence checking is the problem of checking whether two circuit descriptions specify the same behavior.

In order to compare the functionality of two circuits, we need to know the relation between their interfaces. If such a one-to-one correspondence between the primary inputs and outputs of the two circuits which have to be compared is not given, it has to be computed before the test for equivalence can be run. This problem is called *Permutation Independent Boolean Comparison Problem* in literature [106, 108, 109]. We investigate it in Chapter 8.

2.1 Equivalence checking of combinational circuits

Let us assume that the two combinational circuits F and G need to be tested for functional equivalence. When the relation between the interfaces of these two circuits is given, then we have to decide whether the circuits F and G implement the same Boolean function. More formally, the *equivalence checking* problem is defined as follows:

> *Given two representations d_f and d_g of two Boolean functions $f, g : \{0,1\}^n \rightarrow \{0,1\}^m$, decide whether the Boolean functions f and g are equal, i.e., whether $f(\alpha) = g(\alpha)$ holds for all $\alpha \in \{0,1\}^n$.*

In general, this problem is co-NP hard [53]. Just assume that d_f is a Boolean formula and g is the constant Boolean function $\mathbf{0}$. Then, *equivalence checking* means to decide whether d_f describes the constant Boolean function $\mathbf{0}$. This problem is the complemented, well-known SAT problem for Boolean formulae which is known to be NP hard. However, the complexity of *equivalence checking* depends on the representation type chosen. For instance, *equivalence checking* applied to reduced ordered binary decision diagrams (BDDs) [20] or multiplicative binary moment diagrams (*BMDs) [22, 23], which describe pseudo–Boolean functions, can be done in constant time.

One of the basic approaches to prove functional equivalence of two combinational circuits is to transform both circuits into canonical forms. In most applications, these are reduced ordered BDDs or *BMDs since the size of other canonical forms of Boolean functions, as for example minterm forms, are usually much larger than the corresponding reduced ordered BDDs and *BMDs. We discuss this problem in further details in Chapter 3. After the construction of the reduced ordered BDDs or *BMDs of the functions under consideration, the equivalence check can be performed by checking whether the two canonical representations are equal. This can be done in constant time. However, there are some major bottlenecks to this approach which have to be considered. We are going to handle this in detail in Chapter 4. Let us discuss them shortly.

Some functions cannot be efficiently represented by BDDs or *BMDs. Bryant [20] proved that reduced ordered BDDs for integer multipliers are of exponential size. The representational power of *BMDs is too weak to efficiently represent integer division [131]. Obviously, when the functional behavior of a Boolean circuit cannot be efficiently represented by a canonical form, the equivalence problem has to be solved with other approaches. These are automatic test pattern generation methods, SAT solving techniques, or structural equivalence checking. For a description of these methods we refer to Chapter 5 and Chapter 6.

Even if the functional behavior of a Boolean circuit can efficiently be represented by a BDD or a *BMD, the memory requirements may grow exponentially *during* the synthesis of a decision diagram. To overcome this difficulty Hulgaard/Williams/Andersen [64] and Hett [60], for example, introduced a new data structure for representing and manipulating Boolean functions. It is called binary expression diagram (BED). BEDs are an extension of BDDs and they are capable of representing any Boolean circuit in linear space. BEDs can be seen as an intermediate representation between circuits which are compact and BDDs which are canonical. We discuss them in Chapter 4 as well.

Most of the papers published in literature handle the problem of proving that an error-free circuit is error-free. For example, Hamaguchi et.al. [57] experimentally showed that error-free 64-bit integer multipliers of several types can be verified to be error-free in some hours using the backward construction method. But there is no evidence that a *faulty* 64-bit integer multiplier can efficiently be proven to be faulty. Ideas of how to attack this problem have been discussed in [74, 145]. We address them in Chapter 4, too.

A further important tool in formal verification is *Black Box Equivalence Checking* [129, 130] which occurs when the specification is known, but only parts of the implementation have been finished or known. It enables the use of verification techniques in early stages of the design. Design errors can be detected already when only a partial implementation is on hand. The basic idea is to combine parts of the implementation which are not yet finished into black boxes. If the implementation differs from the specification for all possible substitutions of the black boxes, a design error exists. This is the topic of Chapter 7.

2.2 Equivalence checking of sequential circuits

Digital sequential circuits are modelled as finite-state machines (FSM). There is a well-developed theory for analyzing FSMs, including checking their equivalence [75] which is based on a (costly) state space traversal. However, in most cases *equivalence checking* is applied to sequential circuits to show that certain circuit modifications have not altered the functional behavior of the circuit.

Often, the state encoding of the two circuits remains the same. In this special case, it is not always necessary to perform a costly state space traversal. Since the output and state transition functions are combinational circuits, it is sometimes sufficient to compare these functions. For that, the correspondence of the latches of the two sequential circuits to be checked has to be known. Unfortunately, there are several cases in which the correspondence may get lost. During synthesis huge hierarchical designs lead to long hierarchical names for internal state variables. Since most backend tools have severe restrictions on name lengths, the synthesizer shortens these internal names accordingly. In particular, if VHDL generics are involved, the names may be changed such that not even a human is able to establish the correspondence. Here the verifier has to be able to establish the latch correspondence based on the individual output and state transition functions to avoid time consuming state space traversal [118]. In Chapter 10 we are going to introduce some methods to handle this problem.

3. Structure of the book

Let us give a short summary of the contents of this book to help you to find the different topics more easily.

In Part I, we review some basic definitions, notions, and notations that we use throughout the book (Chapter 2). Then we turn our attention to representations of Boolean functions (Chapter 3). We discuss advantages and disadvantages of the different types of representations with respect to formal equivalence checking.

Part II gives a detailed introduction into the principles and methods of *equivalence checking of combinational circuits*. Chapter 4 presents the approach of proving functional equivalence of two combinational circuits by transforming both circuits into canonical forms, normally reduced ordered BDDs or reduced ordered *BMDs [20, 23, 57]. Furthermore, it handles the approach of Hulgaard et.al. [64] and Hett [60] mentioned above and discusses the problem of proving a faulty combinational circuit to be faulty [74, 145]. Chapter 5 is devoted to equivalence checking tools based on automatic test pattern generation and satisfiability of conjunctive normal forms. Since the equivalence checking tools based on canonical forms as well as ATPG- or SAT-based equivalence checking tools often fail when applied to large circuit descriptions, the circuits to be compared have to be simplified then. Approaches which try to simplify the equivalence checking problem by exploiting structural similarities of the circuits to be compared are presented in Chapter 6 [9, 16, 77]. The problem of *Black Box Equivalence Checking* is addressed by our guest authors Bernd

Becker and Christoph Scholl in Chapter 7. Finally, *Permutation Independent Boolean Comparison* is handled in Chapter 8.

In Part III, we are going to give a detailed introduction to *equivalence checking of sequential circuits*. We begin with formal basics in Chapter 9. Furthermore, we are going to dwell on state space traversal, which is the basis of most of the general equivalence checking methods for sequential circuits, in this chapter as well. Chapter 10 deals with the *latch correspondence problem* which is the problem of finding the correspondence between the latches of the sequential circuits under comparison assuming that the state encodings of both circuits are the same.

4. The audience

This book is designed as a textbook. It reviews the basic ideas of current formal equivalence checking methods for combinational and sequential circuits based on canonical representations of Boolean functions, satisfiability solving, automatic test pattern generation, and structural methods. We wish to address a broad audience. *Equivalence Checking of Digital Circuits* is going to serve as

- a textbook for upper-level undergraduate and graduate students in electrical and computer engineering and
- as a reference for circuit design engineers.

Enjoy it!

Contact address: molitor@acm.org

I

FUNDAMENTALS

Since this book is particularly intended to be used as a self-contained textbook, we start with basic definitions and elementary considerations regarding to Boolean functions. The concept plays a central role in formal verification of circuits as Boolean functions are usually used to model combinational circuits. The other central concept, finite state machines which are used to model sequential circuits, is going to be presented in Part III.

After the introduction of the basic notions and notations which we are going to use throughout the book, we will turn our attention to representations of Boolean functions in Chapter 3. We discuss space requirements and operational complexities of the different types of representations which are truth tables, Boolean expressions, cubes, networks, and decision diagrams. Here, we are specially interested in those operations occurring during formal equivalence checking.

Chapter 2

PRELIMINARIES

This chapter briefly reviews the basic concepts of switching theory. This summary is not complete; only those topics are considered which are of relevance for the understanding of later chapters.

1. Basic notations

First of all, we introduce basic notations which we are going to use in the book. We mainly use the notations introduced in [36].

We use braces (i.e., $\{\}$) to denote *unordered* sets and parentheses (i.e., [ ]) to denote *ordered* sets. For example, an unordered set of three elements is denoted by $M = \{a, b, c\}$. The *cardinality* of a set M is the number of its elements. It is denoted by $|M|$. Given a set M, a *partition* of M is a set of disjoint subsets of M whose union is M. For example, $P_M = \{\{b\}, \{a, c\}\}$ is a partition of our example set. Set membership of an element is denoted by $\in$, set inclusion by $\subset$ or $\subseteq$. The symbol $\forall$ is the *universal quantifier*, the symbol $\exists$ the *existential quantifier*. *Implication* is denoted by $\Longrightarrow$ and *co–implication* by $\Longleftrightarrow$. The symbol : means *such that*.

$\mathbb{Z} = \{\ldots, -2, -1, 0, 1, 2, \ldots\}$ denotes the set of integers. $\mathbb{N} = \{1, 2, 3, \ldots\}$ denotes the set of natural numbers. For a fixed $n \geq 1$, $\mathbb{N}_n$ denotes the subset $\{1, 2, 3, \ldots, n\} \subset \mathbb{N}$. With $\mathbb{N}_0$ we denote the set of non-negative integers $\{0, 1, 2, 3, \ldots\}$.

The *Cartesian product* of two sets X and Y, denoted by $X \times Y$, is the set of all ordered pairs (x, y), such that $x \in X$ and $y \in Y$.

A *relation* R between two sets X and Y is a subset of $X \times Y$. We write xRy when $x \in X$, $y \in Y$ and $(x, y) \in R$. An *equivalence relation* is a subset R of $X \times X$ which is *reflexive*, i.e.,

$$\forall x \in X : xRx,$$

symmetric, i.e.,

$$\forall x, y \in X : xRy \Longrightarrow yRx,$$

and *transitive*, i.e.,

$$\forall x, y, z \in X : (xRy \text{ and } yRz) \Longrightarrow xRz.$$

A *partial order* is a relation between X and itself that is reflexive, anti–symmetric, i.e.,

$$\forall x, y \in X : (xRy \text{ and } yRx) \Longrightarrow x = y,$$

and transitive.

A *function (or map)* between two sets X and Y is a relation which has the property that each element of X appears as the first element in one and only one pair of the relation. A function between two sets X and Y is denoted by $f : X \longrightarrow Y$. The sets X and Y are called the *domain* and the *co–domain* of the function, respectively. The set $f(X) = \{f(x) : x \in X\}$ is called the *range* of function f. If each element of the range of a function f has a unique element in the domain of f which maps to it, then we say the function is *injective*. We say a function is *surjective* when the range is equal to the co–domain. Finally, a function is said to be *bijective* if it is both *injective* and *surjective*.

Let M be a finite set, then $\mathcal{P}er(M) = \{\pi;\ \pi : M \to M \text{ is a bijective function}\}$ denotes the set of all permutations of M.

Let A be a finite alphabet, i.e., a finite set of symbols, then $A^\star$ and A^+ denote the set of the *finite sequences* of elements of A and the set of the non-empty finite sequences of elements of A, respectively. The length of a sequence $w \in A^\star$ which is defined to be the number of elements in w is denoted by $|w|$. For all $w \in A^+$ and $1 \leq i \leq j \leq |w|$, the ith element of w is denoted by w_i and the sub-sequence $(w_i, \ldots, w_j)$ of sequence w is denoted by $w_{i \ldots j}$.

2. Boolean algebra

A central mathematical notion in circuit design is the *algebra of logic*, or *Boolean algebra*, which has been introduced by Boole in 1847 [13]. For the definition of Boolean algebras we follow the book of Hachtel and Somenzi [56].

Boolean algebras are based on partially ordered sets.

DEFINITION 2.1 (POSET) *Let M be a set and $R \subseteq M \times M$ a relation. Then (M, R) is called partially ordered set, or poset, if R is a partial order.*

If $(M, \leq)$ is a poset and any two elements $a, b \in M$ have both *meet* and *join*, which we denote by $a \cdot b$ and $a + b$, respectively, then $(M, \leq)$ is a *lattice*.

DEFINITION 2.2 (MEET) *Let $(M, \leq)$ be a poset. An element $m \in M$ is called meet of two elements $a, b \in M$, i.e., $m = a \cdot b$, if*

- *$m \leq a$, $m \leq b$, and*
- *$\forall q \in M : (q \leq a$ and $q \leq b) \Rightarrow q \leq m$*

hold.

DEFINITION 2.3 (JOIN) *Let $(M, \leq)$ be a poset. An element $m \in M$ is called join of two elements $a, b \in M$, i.e., $m = a + b$, if*

- *$a \leq m$, $b \leq m$, and*
- *$\forall q \in M : (a \leq q$ and $b \leq q) \Rightarrow m \leq q$*

hold.

DEFINITION 2.4 (LATTICE) *A poset $(M, \leq)$ is a lattice if meet and join is defined for all elements $a, b \in M$.*

Each finite lattice $(M, \leq)$ meets some properties, namely

LEMMA 2.1 (IDEMPOTENCE)

$$\forall a \in M : a \cdot a = a \text{ and } a + a = a.$$

LEMMA 2.2 (COMMUTATIVITY)

$$\forall a, b \in M : a \cdot b = b \cdot a \text{ and } a + b = b + a.$$

LEMMA 2.3 (ASSOCIATIVITY)

$$\forall a, b, c \in M : (a \cdot b) \cdot c = a \cdot (b \cdot c) \text{ and } (a + b) + c = a + (b + c).$$

LEMMA 2.4 (ABSORPTION)

$$\forall a, b \in M : a \cdot (a + b) = a \text{ and } a + (a \cdot b) = a.$$

LEMMA 2.5 (LAWS OF 0 AND 1)

- $\exists 1 \in M \; \forall a \in M : a \leq 1.$

- $\exists 0 \in M \ \forall a \in M : \ 0 \leq a.$
- $\forall a \in M : \ a \cdot 0 = 0, \ a \cdot 1 = a, \ a + 0 = a,$ and $a + 1 = 1.$

LEMMA 2.6

$$\forall a, b, c \in M : \qquad (a \cdot b) + (a \cdot c) \leq a \cdot (b + c) \text{ and}$$
$$a + (b \cdot c) \leq (a + b) \cdot (a + c).$$

Now, a Boolean algebra is a complemented and distributive lattice.

DEFINITION 2.5 (COMPLEMENTED LATTICE, COMPLEMENT) *A finite lattice $(M, \leq)$ is complemented if for all element $a \in M$ there exists an element b such that $a \cdot b = 0$ and $a + b = 1$. Element b is called complement of a and is denoted by a' or $\neg a$.*

DEFINITION 2.6 (DISTRIBUTIVE LATTICE) *A lattice $(M, \leq)$ is distributive if for all $a, b, c \in M$*

$$(a \cdot b) + (a \cdot c) = a \cdot (b + c) \text{ and } (a + b) \cdot (a + c) = a + (b \cdot c)$$

hold.

DEFINITION 2.7 (BOOLEAN ALGEBRA) *A Boolean algebra is a complemented, distributive lattice.*

As an example, look at the two values to which signals of a circuit are restricted, L and H, or 0 and 1. Combining 0 and 1 by logical-and results in value 0. Now, let us interpret the binary logical-and operation as meet operation. Then $0 \leq 1$ obviously holds and $(\{0, 1\}, \leq)$ is a poset. Furthermore it is easy to prove that this poset is a lattice which is finite, complemented, and distributive. Thus, $(\{0, 1\}, \leq)$ is a Boolean algebra, the so-called *switching algebra*. By this, it is easy to show that the poset $(\{0, 1\}^n, \leq)$ defined by

$$(\alpha_1, \ldots, \alpha_n) \leq (\beta_1, \ldots, \beta_n) :\Longleftrightarrow \forall i \in \mathbb{N}_n \ \alpha_i \leq \beta_i$$

is a Boolean algebra, too. For more details, please see [19] and [99].

3. Boolean functions

In circuit design, Boolean functions play a dominant role. They are based on Boolean n–spaces. A Boolean n–space is a multi–dimensional space spanned by n binary–valued (so called Boolean) variables. It is denoted by $\{0, 1\}^n$.

3.1 Definition and properties

A *completely specified Boolean function* is a mapping between two Boolean spaces and can be defined as follows.

DEFINITION 2.8 (COMPLETELY SPECIFIED BOOLEAN FUNCTION) *Let n and m be two natural numbers and $X_n = \{x_1, \ldots, x_n\}$ a set of n input variables. A function $f : \{0,1\}^n \rightarrow \{0,1\}^m$ which associates every element of $\{0,1\}^n$ with an element of $\{0,1\}^m$ is called m-ary completely specified Boolean function with n input variables or m-ary completely specified Boolean function over X_n. To simplify matters, we simply speak of Boolean functions in most cases. The set X_n is called support of Boolean function f.*

If the order of the variables is important, then we use X_n as a sequence of variables. This will be denoted by the notation $[x_1, \ldots, x_n]$.

For fixed natural numbers n and m, we denote the set $\{f;\ f : \{0,1\}^n \rightarrow \{0,1\}^m\}$ of the m-ary Boolean functions with n inputs by $\mathcal{B}_{n,m}$. If $m = 1$, then we use the notation $\mathcal{B}_n$ instead of $\mathcal{B}_{n,1}$.

Each m-ary Boolean function $f \in \mathcal{B}_{n,m}$ can be interpreted as an m-dimensional vector $(f_1, f_2, \ldots, f_m)$ or set $\{f_1, f_2, \ldots, f_m\}$ of Boolean functions $f_i \in \mathcal{B}_n$.

Let us consider all m-ary Boolean functions with n input variables. Those functions can be ordered. We distinguish the following two ordering relations for Boolean functions.

DEFINITION 2.9 (SMALLER) *A Boolean function $f := (f_1, \ldots, f_m) \in \mathcal{B}_{n,m}$ is less than or equal to a Boolean function $g := (g_1, \ldots, g_m) \in \mathcal{B}_{n,m}$, i.e., $f \leq g$, if for all $i \in \mathbb{N}_m$ and for all assignments $\alpha \in \{0,1\}^n$ the relation $f_i(\alpha) \leq g_i(\alpha)$ holds.*

DEFINITION 2.10 (LEXICOGRAPHICAL SMALLER) *A Boolean function $f \in \mathcal{B}_n$ is lexicographically smaller than or equal to a Boolean function $g \in \mathcal{B}_n$, i.e., $f \leq_{lex} g$, if either $f = g$ or $f(\alpha_1, \ldots, \alpha_n) = 0$ and $g(\alpha_1, \ldots, \alpha_n) = 1$ for the binary vector $(\alpha_1, \ldots, \alpha_n) \in \{0,1\}^n$ with minimal $\sum_{j=1}^{n} \alpha_j 2^{j-1}$ provided that $f(\alpha_1, \ldots, \alpha_n) \neq g(\alpha_1, \ldots, \alpha_n)$.*

For illustration, let us consider the Boolean functions f, g, $h \in \mathcal{B}_2$ defined by

x_2	x_1	f	g	h
0	0	0	0	0
0	1	0	1	1
1	0	0	1	1
1	1	1	1	0

It is easy to check that $f \leq g$, $f \leq_{lex} g$, $f \not\leq h$, and $f \leq_{lex} h$ hold.

Both $(\mathcal{B}_{n,m}, \leq)$ and $(\mathcal{B}_{n,m}, \leq_{lex})$ are posets. $(\mathcal{B}_{n,m}, \leq_{lex})$ is not a complemented lattice in general wheras $(\mathcal{B}_{n,m}, \leq)$ is even a Boolean algebra. Any two Boolean functions f and g of $\mathcal{B}_{n,m}$ have both meet and join with respect to relation $\leq$ which are the logical-and, which we denote by $\cdot$ (or $\wedge$), and the logical-or, which we denote by $+$ (or $\vee$), of f and g, respectively. Thus, $(\mathcal{B}_{n,m}, \leq)$ is a finite lattice and there is a largest element, denoted by $\mathbf{1}$, and a smallest element, denoted by $\mathbf{0}$. Each Boolean function f has a well defined complement f' such that meet of f and f' is $\mathbf{0}$ and join of f and f' is $\mathbf{1}$. Finally, it is easy to prove that the distributivity laws with respect to meet and join hold.

In practice, the functional behavior of a circuit has not to be completely specified. This leads us to the definition of *incompletely specified Boolean functions*. An incompletely specified Boolean function is defined over a subset D of the Boolean n–space $\{0,1\}^n$ only. The elements of D where the function is not defined are called *don´t care* conditions.

DEFINITION 2.11 (INCOMPLETELY SPECIFIED FUNCTION) *Let n be a natural number. A function $f : D \to \{0,1\}$ with $D \subseteq \{0,1\}^n$ is called incompletely specified Boolean function with n inputs. D is the domain of f which is denoted by $domain(f)$. The set $dc(f)$ which contains all elements of $\{0,1\}^n$ not in $domain(f)$ is called don't care set of f. The sets $on(f)$ and $off(f)$ which are called on-set of f and off-set of f, respectively, are defined as*

$$on(f) := \{\alpha \in domain(f); f(\alpha) = 1\}$$

and

$$off(f) := \{\alpha \in domain(f); f(\alpha) = 0\},$$

respectively.

For a fixed subset $D \subseteq \{0,1\}^n$, we denote the set of Boolean functions with domain D by $\mathcal{S}(D)$.

If we consider m-ary incompletely specified Boolean functions with n input variables, then the don't care conditions may differ for each output of the function. Therefore, we often represent them as $f : \{0,1\}^n \longrightarrow \{0,1,*\}^m$, where $*$ represents a *don't care* condition.

If we implement an incompletely specified Boolean function $f \in \mathcal{S}(D)$ by a combinational circuit G, then the circuit G will produce an output for each assignment to the primary inputs, even for those assignments which are in the don't care set of f. Hence, every implementation G of an incompletely specified Boolean function $f \in \mathcal{S}(D)$ for some $D \subseteq \{0,1\}^n$ realizes a (completely specified) Boolean function $g \in \mathcal{B}_n$ which we call extension of function f.

DEFINITION 2.12 (EXTENSION OF FUNCTIONS) *Let f and g be incompletely specified Boolean functions. Function g is an extension of f if and only if $domain(f)$ is a subset of $domain(g)$ and $g(\alpha) = f(\alpha)$ holds for every element $\alpha \in domain(f)$.*

Now we are going to discuss one more property of Boolean functions. For that, let us consider the following. $\mathcal{S}(\{0,1\}^n)$ is the set of incompletely specified Boolean functions with domain $\{0,1\}^n$. Obviously, both $\mathcal{S}(\{0,1\}^n)$ and $\mathcal{B}_n$ describe the same set of functions which is the set of completely specified Boolean functions. In other words, we can consider the set of completely specified Boolean functions as a special case of incompletely specified Boolean functions. Hence, any fact which we are going to discuss for incompletely specified Boolean functions can be applied to completely specified Boolean functions as well. However, note that we assume all functions to be completely specified if not explicitly stated otherwise in the following.

We close this section with further notations which we are going to use in later chapters.

Let $f \in \mathcal{S}(D)$ be an incompletely specified Boolean function with $D \subseteq \{0,1\}^n$. Then, $ON(f)$, $OFF(f)$, and $DC(f)$ denote completely specified Boolean functions of $\mathcal{B}_n$ defined by

$$\forall \alpha \in \{0,1\}^n : ON(f)(\alpha) = 1 \iff \alpha \in on(f),$$

$$\forall \alpha \in \{0,1\}^n : OFF(f)(\alpha) = 1 \iff \alpha \in off(f),$$

and

$$\forall \alpha \in \{0,1\}^n : DC(f)(\alpha) = 1 \iff \alpha \in dc(f),$$

respectively. $|f|$ denotes the number of elements in $on(f)$. This number is called *satisfy count* of function f.

$\boldsymbol{x}_i \in \mathcal{B}_n$ with $i \in \mathbb{N}_n$ denotes the Boolean function defined by

$$\forall (\alpha_1, \alpha_2, \ldots, \alpha_n) \in \{0,1\}^n : \boldsymbol{x}_i(\alpha_1, \alpha_2, \ldots, \alpha_n) = \alpha_i.$$

In other words, we can interpret variable x_i as Boolean function of $\mathcal{B}_n$ which outputs the ith argument.

$\oplus$ denotes the exclusive-or of Boolean functions. For two Boolean functions $f, g \in \mathcal{B}_n$, $f \oplus g$ is defined by

$$f \oplus g = (f \cdot g') + (f' \cdot g).$$

$\overline{\cdot}$, $\overline{+}$, and $\overline{\oplus}$ denote the nand, nor, and equivalence of Boolean functions, respectively. For two Boolean functions $f, g \in \mathcal{B}_n$, these operators are formally defined as $f \overline{\cdot} g = (f \cdot g)'$, $f \overline{+} g = (f + g)'$, and $f \overline{\oplus} g = (f \oplus g)'$, respectively.

3.2 Cofactors

In order to analyze a Boolean function it can be helpful to decompose it with respect to a Boolean variable. In this context, cofactors play an important role.

DEFINITION 2.13 (COFACTOR) *Let $f \in \mathcal{B}_{n,m}$ be a Boolean function. The cofactor of f with respect to $x_i = c$ with $c \in \{0,1\}$ is given by the mapping $f_{x_i=c}$ of $\mathcal{B}_{n,m}$, defined by*

$$\begin{aligned}&\forall \alpha := (\alpha_1, \dots, \alpha_{i-1}, \alpha_i, \alpha_{i+1}, \dots, \alpha_n) \in \{0,1\}^n : \\ &\qquad f_{x_i=c}(\alpha) := f(\alpha_1, \dots, \alpha_{i-1}, c, \alpha_{i+1}, \dots, \alpha_n).\end{aligned}$$

$f_{x_i=0}$ and $f_{x_i=1}$ are called negative and positive cofactor of f with respect to x_i, respectively.

Note that the cofactor $f_{x_i=c}$ of a Boolean function of $\mathcal{B}_{n,m}$ can be interpreted in a natural way as function of $\mathcal{B}_{n-1,m}$ since the cofactor depends on at most $n-1$ variables only.

Often the relation between both cofactors of a function with respect to a variable x_i might be interesting as well. So let us consider them. For the sake of simplicity we consider Boolean functions of $\mathcal{B}_n$ only.

DEFINITION 2.14 (BOOLEAN DIFFERENCE) *Let $f \in \mathcal{B}_n$ be a Boolean function, then $\partial f / \partial x_i$ denotes the Boolean function defined by*

$$\partial f / \partial x_i := f_{x_i=0} \oplus f_{x_i=1}.$$

We call it Boolean difference.

The Boolean difference of a function f with respect to a variable x_i indicates whether x_i is observable at f. When it is zero, then the function does not depend on x_i.

DEFINITION 2.15 (UNIVERSAL AND EXISTENTIAL QUANTIFICATION) *The universal quantification and existential quantification of a Boolean function $f \in \mathcal{B}_n$ with respect to a variable x_i are defined by*

$$(\forall x_i : f) := f_{x_i=0} \cdot f_{x_i=1}$$

and

$$(\exists x_i : f) := f_{x_i=0} + f_{x_i=1},$$

respectively.

The universal quantification of a Boolean function f with respect to a variable x_i represents the part of the domain of f for which both assignments to x_i

evaluate the function to 1, whereas the existential abstraction represents the part of the domain for which at least one assignment to x_i evaluates to 1.

Cofactoring with respect to different variables can by applied successively. It is important to note that cofactoring is commutative, i.e.,

$$(f_{x_i=c_i})_{x_j=c_j} = (f_{x_j=c_j})_{x_i=c_i}.$$

In the following, we use the notation $f_{[x_i,x_j]=(c_i,c_j)}$ to denote the iterated cofactor $(f_{x_i=c_i})_{x_j=c_j}$.

Now let us consider some properties which allow to decompose Boolean functions into simpler subfunctions. These properties were introduced by Boole [14].

THEOREM 2.1 (BOOLE-SHANNON EXPANSION) *Let $f \in \mathcal{B}_n$ be a Boolean function defined over the variable set X_n. Then f can be decomposed in terms of a variable x_i with $i \in \mathbb{N}_n$ as*

$$f = (x_i' \cdot f_{x_i=0}) + (x_i \cdot f_{x_i=1}).$$

This decomposition is called Boole-Shannon expansion.

The proof of the theorem is straightforward since at least one of the terms $x_i \cdot f_{x_i=1}$ and $x_i' \cdot f_{x_i=0}$ have to evaluate to 0 for any assignment of the variables.

THEOREM 2.2 (DAVIO DECOMPOSITION) *Let $f \in \mathcal{B}_n$ be a Boolean function over X_n. Then f can be decomposed in terms of a variable $x_i \in X_n$ as*

$$f = f_{x_i=0} \oplus (x_i \cdot \partial f / \partial x_i)$$

and

$$f = f_{x_i=1} \oplus (x_i' \cdot \partial f / \partial x_i)$$

These decompositions are called positive Davio decomposition and negative Davio decomposition, respectively.

PROOF: Since for each assignment either x_i or x_i' evaluates to 0, we can transform the Boole-Shannon expansion as follows:

$$\begin{aligned}
f &= (x_i \cdot f_{x_i=1}) + (x_i' \cdot f_{x_i=0}) \\
&= (x_i \cdot f_{x_i=1}) \oplus (x_i' \cdot f_{x_i=0}) \\
&= (x_i \cdot f_{x_i=1}) \oplus ((1 \oplus x_i) \cdot f_{x_i=0}) \\
&= (x_i \cdot f_{x_i=1}) \oplus f_{x_i=0} \oplus (x_i \cdot f_{x_i=0}) \\
&= f_{x_i=0} \oplus (x_i \cdot (f_{x_i=1} \oplus f_{x_i=0})).
\end{aligned}$$

Analogously, the second equation of the theorem can be proven. ▪

Last but not least let us mention another fact. The cofactor computation is a special case of the *substitution-by-function* operation which plays an important role during equivalence checking of combinational circuits as well.

DEFINITION 2.16 (SUBSTITUTION BY FUNCTION)
Let $f \in \mathcal{B}_{n,m}$ be a Boolean function defined over the variable set X_n and $g \in \mathcal{B}_n$ a Boolean function which is defined over the same set of variables. Then $f_{x_i=g}$ denotes the Boolean function defined by

$$\forall(\alpha_1, \ldots, \alpha_n) \in \{0,1\}^n : \\ f_{x_i=g}(\alpha_1, \ldots, \alpha_n) = f(\alpha_1, \ldots, \alpha_{i-1}, g(\alpha_1, \ldots, \alpha_n), \alpha_{i+1}, \ldots, \alpha_n).$$

4. Pseudo–Boolean functions

Let us consider another important tool for formal logic verification a little bit closer now. For the purposes of formal verification, m-ary Boolean functions are often considered as integer–valued Boolean functions.

Let $f \in \mathcal{B}_{n,m}$ be an m-ary Boolean function with n input variables. We can associate the mapping

$$F : \{0,1\}^n \rightarrow \mathbb{Z}$$

defined by

$$\forall \alpha \in \{0,1\}^n : \; F(\alpha) := \sum_{i=1}^{m} f_i(\alpha) \cdot 2^{i-1}$$

with f. Note that $\cdot$ denotes integer multiplication here. The iterated sum symbol denotes integer addition.

We call this type of function *pseudo–Boolean function.* They are defined as follows.

DEFINITION 2.17 (PSEUDO–BOOLEAN FUNCTION) *Let n be a natural number. A function $F : \{0,1\}^n \rightarrow \mathbb{Z}$ which associates every element of $\{0,1\}^n$ with an integer is called pseudo–Boolean function. We denote the set of pseudo–Boolean functions with n inputs by $\mathcal{Z}_n$.*

A well-known pseudo–Boolean function is the two's complement decoding function $f : \{0,1\}^n \rightarrow \mathbb{Z}$ defined by

$$f(x_n, x_{n-1}, \ldots, x_1) = \sum_{i=1}^{n-1} x_i \cdot 2^{i-1} - x_n \cdot 2^{n-1}.$$

Similar to Boolean functions, pseudo–Boolean functions can be decomposed with respect to a variable x_i.

DEFINITION 2.18 (COFACTOR OF A PSEUDO–BOOLEAN FUNCTION) *Let $f \in \mathcal{Z}_n$ be a pseudo–Boolean function. The cofactor of f with respect to $x_i = c$ with $c \in \{0, 1\}$ is given by the mapping $f_{x_i=c}$ of $\mathcal{Z}_n$, defined by*

$$\forall \alpha := (\alpha_1, \ldots, \alpha_{i-1}, \alpha_i, \alpha_{i+1}, \ldots, \alpha_n) \in \{0,1\}^n :$$
$$f_{x_i=c}(\alpha) := f(\alpha_1, \ldots, \alpha_{i-1}, c, \alpha_{i+1}, \ldots, \alpha_n).$$

$f_{x_i=0}$ and $f_{x_i=1}$ are called negative and positive cofactor of f with respect to x_i, respectively.

Again, the cofactor $f_{x_i=c}$ of a pseudo–Boolean function of $\mathcal{Z}_n$ can be interpreted as function of $\mathcal{Z}_{n-1}$ since the cofactor depends on at most $n - 1$ variables only. Cofactoring of pseudo–Boolean functions has the same properties as cofactoring for Boolean functions, i.e., it can be applied successively and it is commutative. Furthermore, the substitution–by–function operation for pseudo–Boolean functions is defined in the same way as this for Boolean functions. Hence, we would like to refer to the previous section about cofactoring for Boolean functions.

A frequently-used relation between both cofactors of a pseudo–Boolean function with respect to a variable x_i is defined as follows.

DEFINITION 2.19 (LINEAR MOMENT)
Let $f \in \mathcal{Z}_n$ be a pseudo–Boolean function, then $\partial f / \partial x_i$ denotes the pseudo–Boolean function defined by

$$\partial f / \partial x_i \; := \; f_{x_i=1} - f_{x_i=0}$$

where $-$ denotes integer subtraction. Function $\partial f / \partial x_i$ is called linear moment of f with respect to x_i and is often denoted by $f_{\dot{x}_i}$ in literature.

Moreover, there is a decomposition theorem for pseudo–Boolean functions as well. It is due to Bryant and Chen [20, 23].

THEOREM 2.3 (MOMENT DECOMPOSITION)
Let $f \in \mathcal{Z}_n$ be a pseudo–Boolean function defined over the variable set X_n. Then f can be decomposed in terms of a variable x_i with $i \in \mathbb{N}_n$ as

$$f = f_{x_i=0} + x_i \cdot f_{\dot{x}_i},$$

where $+$ and $\cdot$ denote integer addition and integer multiplication, respectively. The decomposition is called moment decomposition of f with respect to variable x_i. In this context, function $f_{x_i=0}$ is called constant moment of f with respect to variable x_i.

PROOF: For any assignment α of the variables either x_i or x_i' evaluates to 0. Thus the equation

$$f = x_i \cdot f_{x_i=1} + x_i' \cdot f_{x_i=0}$$

holds where $+$ and $\cdot$ denote integer addition and integer multiplication, respectively. As the pseudo–Boolean function $1 - x_i$ represents x_i', the above equation is equivalent to

$$f = (x_i \cdot f_{x_i=1}) + (1 - x_i) \cdot f_{x_i=0}$$

and the theorem immediately follows. ∎

Chapter 3

REPRESENTATION OF BOOLEAN AND PSEUDO BOOLEAN FUNCTIONS

The efficiency of algorithms for specific problems highly depends on an appropriate representation of the objects which have to be manipulated. In this chapter, we review and analyze the different representations of Boolean functions with respect to their applicability in the context of formal verification. So we are able to better understand why some data structures are used and why others are not used.

Following [76], we start with a formal definition of representations of Boolean or pseudo–Boolean functions. For a fixed n and m, let $\mathcal{F}_{n,m}$ be either the set $\mathcal{B}_{n,m}$ of m-ary Boolean functions with n input variables or the set $\mathcal{Z}_n$ of pseudo–Boolean functions with n input variables.

DEFINITION 3.1 (REPRESENTATION OF FUNCTIONS) *Let $\mathcal{R}_{n,m}$ be a finite set and ϕ be a mapping which uniquely associates every element $r \in \mathcal{R}_{n,m}$ with a function of $\mathcal{F}_{n,m}$, i.e., $\phi : \mathcal{R}_{n,m} \to \mathcal{F}_{n,m}$. Then, $(\mathcal{R}_{n,m}, \phi)$ is called representation of the functions of $\mathcal{F}_{n,m}$. The mapping ϕ is called interpretation of $\mathcal{R}_{n,m}$. If $\phi(r) = f$ holds, then r is called representation of f.*

1. Desirable properties of representations

Representations of Boolean and pseudo–Boolean functions should have some special properties in order to be applicable for formal verification.

DEFINITION 3.2 *A representation $(\mathcal{R}_{n,m}, \phi)$ of functions of $\mathcal{F}_{n,m}$ is called*

complete *if ϕ is surjective, i.e., if for every function $f \in \mathcal{F}_{n,m}$ there is an element $r \in \mathcal{R}_{n,m}$ such that r is a representation of f. If $(\mathcal{R}_{n,m}, \phi)$ is complete, $\mathcal{R}_{n,m}$ is said to be a representation of $\mathcal{F}_{n,m}$.*

unique *if ϕ is injective, i.e., if $\phi(r) \neq \phi(q)$ holds for any two representations $r, q \in \mathcal{R}_{n,m}$ with $r \neq q$.*

canonical *if ϕ is bijective, i.e., if the representation is complete and unique.*

Uniqueness is a very important property with respect to equivalence checking as whenever a representation $(\mathcal{R}_{n,m}, \phi)$ is unique then for each $s, t \in \mathcal{R}_{n,m}$ the represented functions $\phi(s)$ and $\phi(t)$ are equal if and only if the representations s and t are equal.

Of course, uniqueness is not quite enough. The data structure has to allow efficient implementations of the essential operations needed during formal equivalence checking. The essential operations are

the **unary** and **binary Boolean operations,** that is, given two representations s and t of Boolean functions of $\mathcal{B}_{n,m}$, compute the representations u of $\phi(s) \cdot \phi(t)$, v of $\phi(s) + \phi(t)$, and w of $\phi(s)'$. These operations are needed to synthesize a representation of the function realized by a given combinational circuit.

cofactor computation with respect to a variable x_i, i.e., given a representation s of a Boolean or pseudo–Boolean function, compute representations u and v of $\phi(s)_{x_i=0}$ and $\phi(s)_{x_i=1}$, respectively. As already mentioned, cofactor computation is applied during the analysis of Boolean functions.

substitution by functions with respect to a variable x_i, i.e., given two representations s and t, compute a representation u of $\phi(s)_{x_i=\phi(t)}$.

tautology check and **satisfiability check,** i.e., given a representation s of a Boolean function, decide whether $\phi(s) = \mathbf{1}$ and $\phi(s) \neq \mathbf{0}$, respectively. It is easy to see that checking whether a given Boolean function is a tautology or whether it is satisfiable is a special case of the equivalence checking problem.

satisfy count, i.e., given a representation s of a Boolean function of $\mathcal{B}_n$, compute $|\phi(s)|$ of $\phi(s)$. Note that the satisfy count can be used as a filter during equivalence checking. Given two representations s, t of Boolean functions of $\mathcal{B}_n$, if $|\phi(s)| \neq |\phi(t)|$ then $\phi(s) \neq \phi(t)$ holds.

There are many other operations which can be efficiently implemented using the operations listed above. We do not go into details here.

In the following, the elements of $\mathcal{R}_{n,m}$ are always going to be considered as a data structure in a computer. Let $c_{size}(r)$ be the size of the data structure of element $r \in \mathcal{R}_{n,m}$, i.e., the number of bits necessary to represent r in a computer. Then, the complexity of a Boolean or pseudo–Boolean function with respect to $(\mathcal{R}_{n,m}, \phi)$ can be defined as

$$c_{size}^{\mathcal{R}_{n,m}}(f) := min\{c_{size}(r);\ r \in \mathcal{R}_{n,m} \text{ and } \phi(r) = f\}.$$

We define $c_{size}^{\mathcal{R}_{n,m}}(f) = +\infty$ for those functions f which have no representations in $\mathcal{R}_{n,m}$.

For practical considerations we have to accept the following fact.

THEOREM 3.1 *Let $(\mathcal{R}_{n,1}, \phi)$ be a complete representation of $\mathcal{B}_n$ and let $p < 2^n$ be a natural number. More than $2^{2^n} \cdot (1 - 2^{p-2^n})$ Boolean functions of $\mathcal{B}_n$ have complexities which are larger than or equal p.*

PROOF: The cardinality of the set $\{r;\ r \in \mathcal{R}_{n,1} \text{ and } c_{size}(r) < p\}$ is less than 2^p as c_{size} denotes the size of r in bits. Thus, less than 2^p Boolean functions of $\mathcal{B}_n$ have complexities which are less than p. It immediately follows that there are more than $2^{2^n} - 2^p$ Boolean functions f of $\mathcal{B}_n$ with $c_{size}^{\mathcal{R}_{n,1}}(f) \geq p$. ■

In other words, there is not any complete representation $(\mathcal{R}_{n,1}, \phi)$ of $\mathcal{B}_n$ with workable size complexity.

2. Traditional representations of Boolean functions

Let us review the most important traditional forms of representations of Boolean functions which are truth tables, Boolean networks, Boolean formulae, and sum of products.

2.1 Truth tables

The apparent way of representing a Boolean function is to use the *truth table* of f. The truth table of a Boolean function f is a complete listing of the function values starting with the function value $f(0, \ldots, 0)$ up to the function value $f(1, \ldots, 1)$. Usually, the truth table of a Boolean function $f \in \mathcal{B}_{n,m}$ is coded as the m-dimensional Boolean vector $F[0 : 2^n - 1]$ with

$$\forall \alpha := (\alpha_1, \ldots, \alpha_n) \in \{0,1\}^n : \ f(\alpha) = F\left[\sum_{i=1}^{n} \alpha_i \cdot 2^{i-1}\right].$$

For illustration, let us consider a Boolean function $f \in \mathcal{B}_{3,2}$. Assume that the first output y_1 of f evaluates to 1, if both input x_1 and input x_2 are set to 1, and the second output y_2 of f is 1, if y_1 is 1 or input x_3 is assigned value 1. This Boolean function can be represented by the truth table F

β	$F[\beta]$
0	(0,0)
1	(0,0)
2	(0,0)
3	(1,1)
4	(0,1)
5	(0,1)
6	(0,1)
7	(1,1)

where β denotes the index of the Boolean vector.

Truth tables have very nice properties. They are a canonical representation of Boolean functions. In particular, the equivalence check can be done in linear time with respect to the size of the representation. The time complexity of the remaining operations listed above is linear, as well. Thus, truth tables are an ideal representation of Boolean functions as long as the number n of inputs is not too large, as the space complexity is exponential in n. Indeed, truth tables are unbeatable for Boolean functions $f \in \mathcal{B}_n$ with $n \leq c$ for some constant $c \in \{8, 9, 10, 11\}$.

2.2 Boolean networks

Basically, Boolean networks are labelled directed acyclic graphs. In contrast to truth tables which represent the pure functional behavior of a circuit, Boolean networks are structural representations of circuits, i.e., they are much more in accordance with the intuitive notion of combinational circuits.

In the following, let $\Omega \subseteq \bigcup_{i \in \mathbb{N}} \mathcal{B}_i$ be a finite subset of single-output Boolean functions which constitutes the *cell library* we use for our circuits. Furthermore, we denote the input pads of the network with $IPAD$ and the output pads with $OPAD$.

DEFINITION 3.3 (BOOLEAN NETWORK [128])
Let $T = \Omega \cup \{IPAD, OPAD\}$. A Boolean network $\mathcal{N}$ over Ω with n inputs and m outputs is a tuple $(G, type, pe, pa)$ with

- *$G = (V, E)$ is a directed acyclic graph which is node oriented, i.e., for all nodes $v \in V$ there is an order for the incoming edges. More exactly, V is the set of nodes (which represent the active elements of the circuit) and E is the set of edges (the signals of the circuit). Source and target of an edge are given by functions $source : E \to V$ and $target : E \to V$. The indegree of a node $v \in V$ is given by the mapping $indeg : V \to \mathbb{N}_0$ which is defined by $indeg(v) = |\{e \in E;\ target(e) = v\}|$. Accordingly, the fanout of v is given by $fanout : V \to \mathbb{N}_0$ with $fanout(v) = |\{e \in E;\ source(e) = v\}|$.*

The node orientations of G are given by a one-to-one mapping I. $I(v, i)$ *denotes the* i*th incoming edge of node* $v \in V$ *for* $i \in \mathbb{N}_{indeg(v)}$.

- $type : V \to T$ *is a mapping which assigns a cell of* T *to each node. There are exactly* n *nodes of type* $IPAD$ *and* m *nodes of type* $OPAD$*. If* $type(v)$ *is a cell of cell library* Ω*, then* $type(v) \in \mathcal{B}_{indeg(v)}$*. If* $type(v) = IPAD$*, then* v *does not have any incoming edge, i.e.,* $indeg(v) = 0$*. If* $type(v) = OPAD$*, then* $indeg(v) = 1$ *and* $fanout(v) = 0$.
- *The mappings* $pe : \mathbb{N}_n \to \{v \in V;\ type(v) = IPAD\}$ *and* $pa : \mathbb{N}_m \to \{v \in V;\ type(v) = OPAD\}$ *are one-to-one functions which give the orders for the* $IPAD$*- and* $OPAD$*-nodes.* $pe(i)$ *and* $pa(j)$ *are called* i*th primary input and* j*th primary output of* $\mathcal{N}$*, respectively.*

The interpretation $\phi(\mathcal{N})$ of a Boolean network $\mathcal{N}$ with n inputs and m outputs can be defined straightforward. We associate every edge e of the Boolean network with the Boolean function f_e which is computed at this edge. Assume that edge e is driven by node v, i.e., $e = (v, w)$ for some node w. If v is the ith primary input then f_e is the Boolean function $\boldsymbol{x}_i$. If v is an inner node, i.e., $type(v) = g$ for some cell $g \in \Omega$, we consider the incoming edges $e_1, \ldots, e_{indeg(v)}$ of v. Assume that $e_k = I(v, k)$ for $k = 1, \ldots, indeg(v)$. Then the Boolean function f_e computed by edge e is given by

$$f_e = g \circ (f_{e_1} \times \ldots \times f_{e_{indeg(v)}}),$$

where $\circ$ and $\times$ denote composition and parallel execution of functions, respectively. Now, consider the primary outputs of $\mathcal{N}$. Let y_j be the incoming edge of the jth primary output. Then, Boolean network $\mathcal{N}$ realizes the m-ary Boolean function

$$\phi(\mathcal{N}) = f_{y_1} \times \ldots \times f_{y_m}$$

which is a Boolean function with n inputs.

For illustration, see Figure 3.1 which shows a Boolean network which realizes the Boolean function already used for illustration in Section 2.1. The input pads (v_1, v_2, v_3) and output pads (v_6, v_7) are represented by the grey nodes and black nodes, respectively.

All the synthesis operations, i.e., the unary and binary Boolean operations, the cofactor computation, and the substitution by functions can be done very efficiently. However, Boolean networks are not a unique representation of Boolean functions. Furthermore, it is well known from complexity theory that satisfiability, tautology, and equivalence check applied to Boolean networks are NP-hard problems and co-NP-hard problems, respectively [53]. Therefore, Boolean networks are not suitable for formal logic verification, in general.

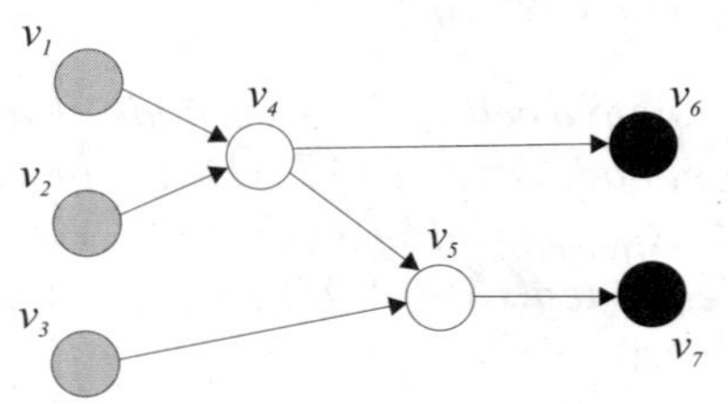

V	type	pe^{-1}	pa^{-1}
v_1	IPAD	1	
v_2	IPAD	2	
v_3	IPAD	3	
v_4	AND		
v_5	OR		
v_6	OPAD		1
v_7	OPAD		2

Figure 3.1. Boolean Network

2.3 Boolean formulae

A very common method to represent Boolean functions is to use Boolean formulae.

DEFINITION 3.4 (BOOLEAN FORMULAE) *The set $\mathcal{BF}(X_n)$ of Boolean formulae over X_n is defined as follows.*

- *The symbols $x_1, \ldots, x_n$, 0, and 1 are Boolean formulae over X_n.*
- *If $w_1, \ldots, w_k$ are Boolean formulae over X_n, then so are the conjunction $(w_1) \cdot \ldots \cdot (w_k)$ and the disjunction $(w_1) + \ldots + (w_k)$.*
- *If w is a Boolean formula over X_n, then $(w)'$ is a Boolean formula over X_n. It is called complement of w.*
- *A string is a Boolean formula over X_n if and only if it can be derived by applying the previous rules a limited number of times.*

To simplify matters, we drop most parentheses by assuming that $\cdot$ takes precedence over $+$. Sometimes we also drop $\cdot$ when no ambiguity arises. Furthermore we use ${x_i}'$ instead of $(x_i)'$, in the following. Sometimes, we use x_i^0 and x_1^1 to denote ${x_i}'$ and x_i, respectively.

The interpretation $\phi : \mathcal{BF}(X_n) \to \mathcal{B}_n$ of Boolean formulae is a homomorphism defined by

- $\phi(x_i) = \boldsymbol{x}_i$,
- $\phi(0) = \mathbf{0}$,
- $\phi(1) = \mathbf{1}$,
- $\phi((w_1) \cdot \ldots \cdot (w_k)) = \phi(w_1) \cdot \ldots \cdot \phi(w_k)$,

- $\phi((w_1) + \ldots + (w_k)) = \phi(w_1) + \ldots + \phi(w_k)$, and
- $\phi((w)') = \phi(w)'$.

We say that the Boolean formula w is a Boolean formula of a Boolean function $f \in \mathcal{B}_n$ if $\phi(w) = f$ holds.

Boolean formulae are usually coded by syntax-trees with respect to the context-free grammar [2]

start symbol: E

<E> → <E>+<T> | <T>

<T> → <T>·<F> | <F>

<F> → <F>′ | (<E>) | 0 | 1 | x_1 | ... | x_n.

For illustration we refer to Figure 3.2. More details can be found in [2].

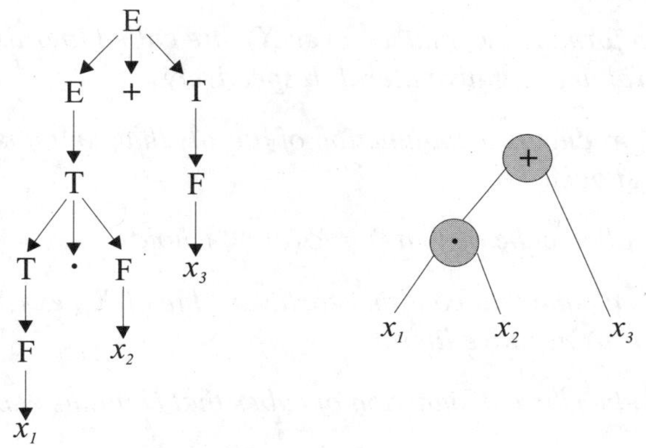

Figure 3.2. At the left: Derivation of the Boolean formula $(x_1 \cdot x_2) + x_3$ with respect to the context-free grammar from above. At the right: the corresponding syntax-tree.

$(\mathcal{BF}(X_n), \phi)$ is a complete representation of $\mathcal{B}_n$. The unary and binary Boolean operations can be done in constant time. Just insert a new root whose outgoing edges point to the operands. Cofactoring can be performed in linear time with respect to the size of the syntax-tree. Let w be a syntax-tree which represents the Boolean function $f \in \mathcal{B}_n$, then a syntax-tree for $f_{x_i=1}$ for some variable x_i can be computed by replacing the leaves marked with symbol x_i by leaves marked with symbol 1. To compute $f_{x_i=0}$, just replace leaves marked with symbol x_i by leaves marked with symbol 0. In the same manner, $f_{x_i=g}$ can be computed for some Boolean function g. Just replace the leaves labelled with x_i by the given syntax-tree of g.

Unfortunately, Boolean formulae are not a unique representation of Boolean functions. In particular, it is easy to see that there exist an infinite number of Boolean formulae which represent the same function, as $\phi(w \cdot w) = \phi(w)$ holds for every Boolean formula $w \in \mathcal{BF}(X_n)$. Worst of all, the decision problem as to whether two given Boolean formulae represent the same Boolean function is co-NP-hard as the well-known satisfiability problem which is the decision problem as to whether a given Boolean formula differs from the constant Boolean function $\mathbf{0}$ is NP-hard. Thus, the decision problem as to whether the satisfy count of a given Boolean formula is larger than 0 is NP-hard, too.

2.4 Sums of products

We can impose restrictions on the form of formulae so that there is only one formula for each Boolean function.

DEFINITION 3.5 (SUM OF PRODUCTS, MINTERM FORM) *Let $f \in \mathcal{B}_n$ be a Boolean function over X_n.*

- *The Boolean formulae x_i and $x_i{}'$ over X_n are called literals which are the positive literal and negative literal, respectively.*
- *A cube or a product is a conjunction of literals that contains each variable of X_n at most once.*
- *A cube c is called cube of f, if $\mathbf{0} \neq \phi(c) \leq f$ holds.*
- *A minterm is a cube that contains each variable of X_n exactly once, either as positive or as negative literal.*
- *A sum of products is a disjunction of cubes that contains each cube at most once.*
- *The sum of products of a Boolean function $f \in \mathcal{B}_n$ which consists of the minterms of f is called minterm form of f. We call the set of the minterms of a Boolean function f satisfy set of f.*

As illustration, let us take the Boolean function $y_2 \in \mathcal{B}_3$ already used in Section 2.1 once more. Obviously, the Boolean formula

$$(x_1 \cdot x_2 \cdot x_3) + (x_1 \cdot x_2 \cdot x_3{}') + (x_1{}' \cdot x_2{}' \cdot x_3) + (x_1 \cdot x_2{}' \cdot x_3) + (x_1{}' \cdot x_2 \cdot x_3)$$

is a minterm form of that Boolean function. It is unique up to permutation of the minterms. The Boolean formula

$$(x_1 \cdot x_2) + x_3$$

is a sum of products of the same Boolean function but not a minterm form.

Since the set of minterms of a Boolean function f is unique, minterm forms are a canonical representation of Boolean functions. Equivalence check, tautology check, and satisfiability check can be done in polynomial time with respect to the size of the minterm forms. Furthermore, the satisfy count of a minterm form can also be computed in linear time. Given the minterm forms of two Boolean functions $f, g \in \mathcal{B}_n$, the minterm forms of $f \cdot g$, $f + g$, and $f_{x_i=g}$ for some variable x_i of f, respectively, can be computed in polynomial time, too. Complementing a minterm form may be a very time consuming operation. Just try to explicitly construct the minterm form of $\phi((x_1 \cdot x_2 \cdot \ldots \cdot x_n)')$.

However, since the size of the minterm forms of most Boolean functions occurring in applications is exponential in the number of variables minterm forms are not used in practice. Indeed, the size of the minterm form of a Boolean function f is proportional to the satisfy count of f which is exponential in the number of variables, in general.

In many applications sum of products are used as representations of Boolean functions as sums of products often are much smaller than minterm forms. Nevertheless, many Boolean functions, for example n-bit binary addition and parity with n input variables, do not have sums of products of practicable size. The smallest sums of products of these Boolean functions are exponential in n.

The satisfiability problem applied to sums of products can be solved in a very easy way. Just take one cube c of the given sum of products and assign Boolean values to the variables occurring in c accordingly. However, as the cubes of a sum of products do not have to be disjoint, tautology check and satisfy count computation cannot be efficiently computed if sums of products are used. Furthermore, complementing sums of products is very time consuming, as well. For example, consider the sum of products $(x_1 \cdot x_2) + (x_3 \cdot x_4) + \ldots + (x_{n-1} \cdot x_n)$ and complement it.

3. Binary Decision Diagrams

Binary decision diagrams (BDDs) as a data structure for representation of Boolean functions were first introduced by Lee [88] and further popularized by Akers [3] and Moret [111]. As BDDs play a dominant role in today's formal verification tools, we will discuss BDDs in detail. This and the next section will follow the paper of Drechsler and Sieling [47], in part.

3.1 Basic definitions

DEFINITION 3.6 (BINARY DECISION DIAGRAM) *A binary decision diagram (BDD) over X_n is a directed acyclic graph $G = (V, E, index, value)$ with a set of nodes V containing two types of nodes, internal and terminal nodes.*

An internal node v has as label a variable $index(v) \in X_n$ and two children $low(v), high(v) \in V$. The edges $(v, low(v))$ and $(v, high(v))$ are called low-edge, or 0-edge, and high-edge, or 1-edge, of v, respectively. The nodes $low(v)$ and $high(v)$ are called low-successor, or 0-successor, and high-successor, or 1-successor, of v, respectively. A terminal node v is labelled with a value $value(v) \in \{0, 1\}$ and has no outgoing edge.

The size $|G|$ of a BDD G is defined as the number of internal nodes of G.

BDDs over X_n can be used to compute Boolean functions of $\mathcal{B}_n$ in the following way. Each input $\alpha = (\alpha_1, \ldots, \alpha_n) \in \{0, 1\}^n$ defines a computation path through the BDD that starts at some distinguished node, the *root* of the corresponding Boolean function. If the path reaches an internal node w that is labelled with x_i, it follows the path $low(w)$ if and only if $\alpha_i = 0$, and it follows the path $high(w)$ if and only if $\alpha_i = 1$. On all paths a terminal node is reached since a BDD is finite, directed, and acyclic. The label of the terminal node determines the return value of the BDD on input α. More formally, we can define the Boolean function corresponding to a BDD and a distinguished node v, recursively.

DEFINITION 3.7 (INTERPRETATION OF BINARY DECISION DIAGRAMS) *Let $G = (V, E, index, value)$ be a BDD over n variables $x_1, \ldots, x_n$ and $v \in V$ a node of BDD G. The Boolean function $\phi(v) \in \mathcal{B}_n$ which is computed at node v is recursively defined as follows.*

- *If v is a terminal node and $value(v) = 1$ then $\phi(v) = \mathbf{1}$.*
- *If v is a terminal node and $value(v) = 0$ then $\phi(v) = \mathbf{0}$.*
- *If v is an internal node and $index(v) = x_i$ then*

$$\phi(v) = \left(x_i{}' \cdot \phi(low(v))\right) + \left(x_i \cdot \phi(high(v))\right).$$

The variable x_i is called decision variable for v.

Figure 3.3 which shows a BDD over X_4 illustrates the definition. The internal nodes which are represented by circles are labelled with x_2, x_3, and x_4. Low-edges are represented by dashed lines. High-edges are represented by solid lines. All the edges are directed from top to bottom. Quadrangles represent terminal nodes. Terminal nodes are labelled with 0 or 1 and represent the Boolean functions $\mathbf{0}$ and $\mathbf{1}$, respectively. Every node v of a BDD represents a Boolean function $\phi(v) \in \mathcal{B}_4$. The left upper internal node and right upper internal node of the BDD represent the Boolean functions described by

$$x_4 + (x_2 \cdot x_3 \cdot x_4{}')$$

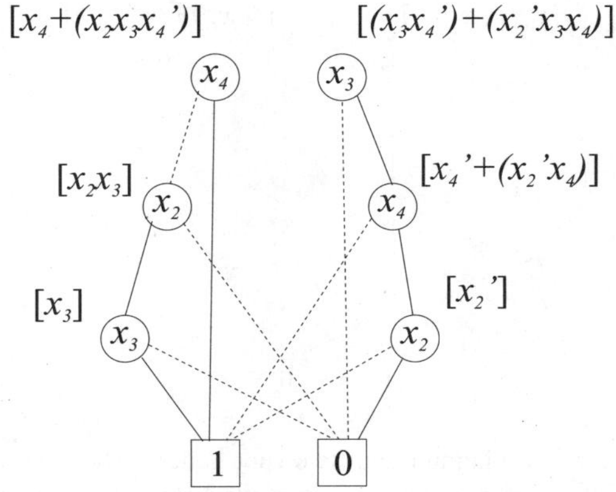

Figure 3.3. Binary decision diagram. The dashed and solid lines denote low-edges and high-edges, respectively. All edges are directed from top to bottom.

and

$$(x_3 \cdot x_4') + (x_2' \cdot x_3 \cdot x_4),$$

respectively. Note that the Boolean functions represented by the BDD are independent of variable x_1 as there is no internal node labelled with x_1.

Notation *When we talk about a BDD F of an m-ary Boolean function $f \in \mathcal{B}_{n,m}$ in the following, we mean a BDD which contains m distinguished nodes $r_{F,1}, \ldots, r_{F,m}$ with $\phi(r_{F,i}) = f_i$ for each $i \in \mathbb{N}_m$. We call them root nodes in the following. In the illustrations, we will mark them with pointers (see Figure 3.4). If there is only one root node r_F, then r_F does not have any incoming edge and we use the notation $\phi(F) = f$.*

Because of the Boole-Shannon expansion, it is easy to see that for each Boolean function $f \in \mathcal{B}_n$ there exists a BDD F with root node r_F representing f, i.e., $\phi(r_F) = f$. Thus, BDDs over X_n are a complete representation of $\mathcal{B}_n$. Allowing more than one distinguished node in the BDDs results in representations of m-ary Boolean functions. More formally, for fixed natural numbers n and m, BDDs over X_n with m root nodes are a complete representation of $\mathcal{B}_{n,m}$.

Unfortunately, an efficient manipulation of BDDs is not possible for the general BDDs defined in Definition 3.7. Therefore, we need restrictions on the structure of the BDDs which were introduced by Bryant [20].

DEFINITION 3.8 (VARIABLE ORDER) *Let $\pi \in \mathcal{P}er(\mathbb{N}_n)$ be any permutation of set $\mathbb{N}_n$. Permutation π induces a linear order $<$ on X_n which is defined by $x_{\pi(i)} < x_{\pi(j)}$ if and only if $i < j$. We say that the linear order $<$ is a variable order.*

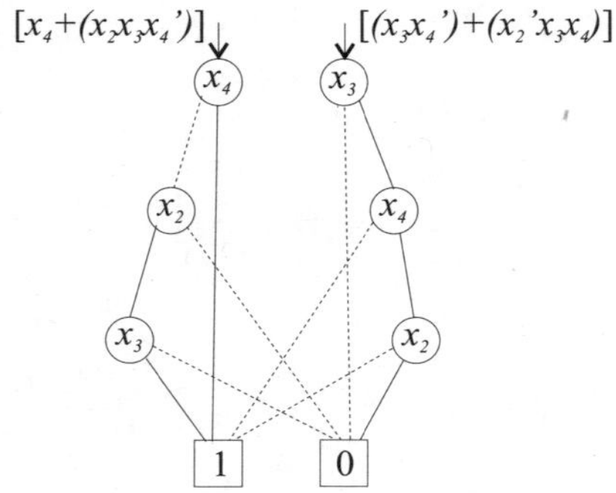

Figure 3.4. Binary decision diagram with two root nodes. The corresponding pointers are given by the arrows without source node. It represents a Boolean function $f \in \mathcal{B}_{4,2}$ described by the two Boolean formulae $x_4 + (x_2 \cdot x_3 \cdot x_4')$ and $(x_3 \cdot x_4') + (x_2' \cdot x_3 \cdot x_4)$.

DEFINITION 3.9 (REDUCED ORDERED BINARY DECISION DIAGRAMS) *A BDD $G = (V, E, index, label)$ is ordered if there exists a linear order $<$ of the variables of G such that $index(u) < index(v)$ holds for each edge $(u, v) \in E$ with v being an internal node. If the BDD G is ordered, then π_G and $<_G$ denote the corresponding permutation and variable order, respectively.*

An ordered BDD is reduced if there are no two different nodes v_1 and v_2 with $\phi(v_1) = \phi(v_2)$.

A BDD which is ordered and reduced is called reduced ordered binary decision diagram ROBDD.

Figure 3.5 illustrates the last definitions. The ordered binary decision diagram which is defined over the variables x_1, x_2, x_3, and x_4 and is ordered according to $x_2 < x_1 < x_3 < x_4$ represents the Boolean function $f \in \mathcal{B}_4$ described by the Boolean formula $(x_1 \cdot x_3 \cdot x_4) + x_2$ – which is equivalent to the Boolean formula $(x_1 \cdot x_2' \cdot x_3 \cdot x_4) + x_2$.

As shown in Figure 3.5, the internal nodes of an ordered binary decision diagram $G = (V, E, index, value)$ over X_n can be graphically partitioned in $n + 1$ levels. For $j \in \mathbb{N}_n$, the jth level consists of those nodes labelled with $x_{\pi_G(j)}$. The terminal nodes are at level $n + 1$. By this, we obtain a mapping $level_G : V \to \mathbb{N}_{n+1}$ which uniquely associates each node of V with its level. In the following, let $level_G^{-1}(j)$ denote the nodes of the jth level of G.

Note that the binary decision diagram shown in Figure 3.5 is not reduced. The three left nodes labelled with x_4 represent the same Boolean function, namely the constant Boolean function **0**. The left node labelled with x_3 also describes the constant Boolean function **0**. Thus, these four nodes can be identified with

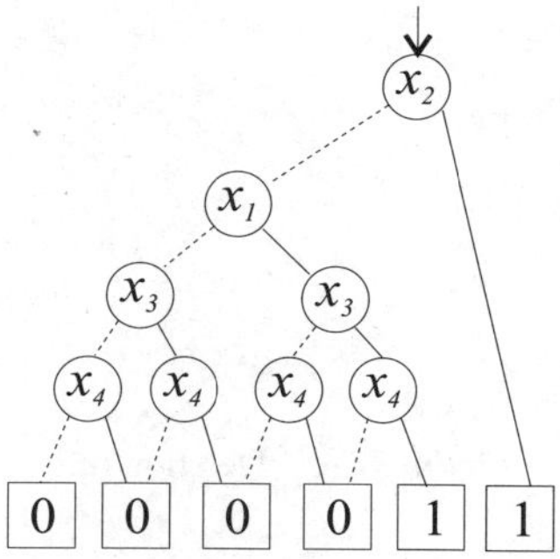

Figure 3.5. Ordered binary decision diagram. The edges are directed from top to bottom.

the terminal nodes labelled with 0 and we obtain the ordered binary decision diagram shown in Figure 3.6 which is reduced.

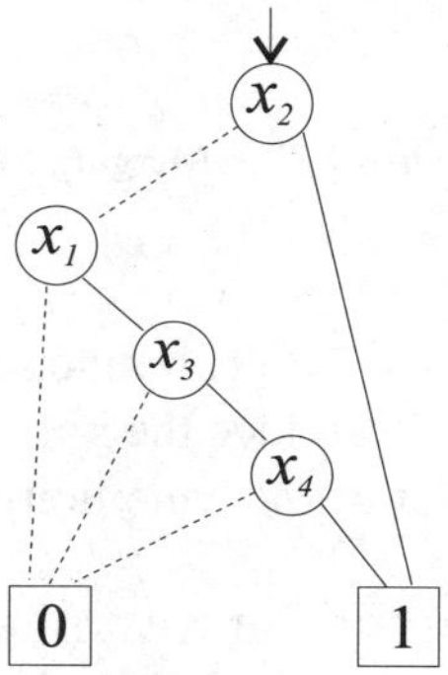

Figure 3.6. ROBDD of the Boolean function $f \in \mathcal{B}_4$ given by $(x_1 x_3 x_4) + x_2$.

The reduction of ordered binary decision diagrams is based on only two local reduction rules. The main idea of the *reduction rules* is to remove redundancies from the ordered binary decision diagram, namely superfluous tests of variables and tests that are represented more than once. Both reduction rules are shown in Figures 3.7a and 3.7b. The reduction rule of Figure 3.7a can be applied to nodes v for which both outgoing edges lead to the same node w. It is obvious that we can redirect all edges leading to v to the node w and that we can delete v afterwards without changing the interpretation of the ordered binary decision diagram. This reduction rule is called *deletion rule*. The reduction rule shown in Figure 3.7b which is called *merging rule* is applicable if there are nodes v_1 and v_2 with the same label, the same low-successor, and the same high-successor. Bryant [20] proved that using these two reduction rules is sufficient to obtain a

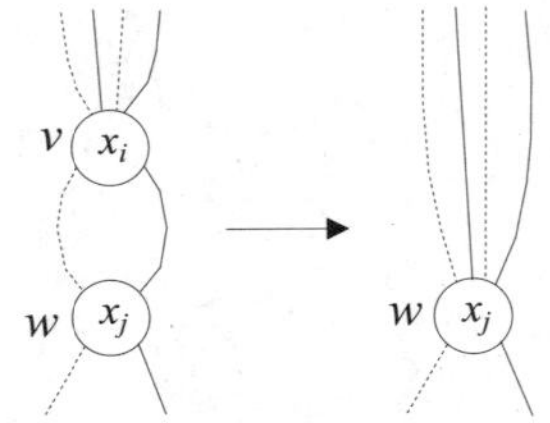

Figure 3.7a. Deletion rule

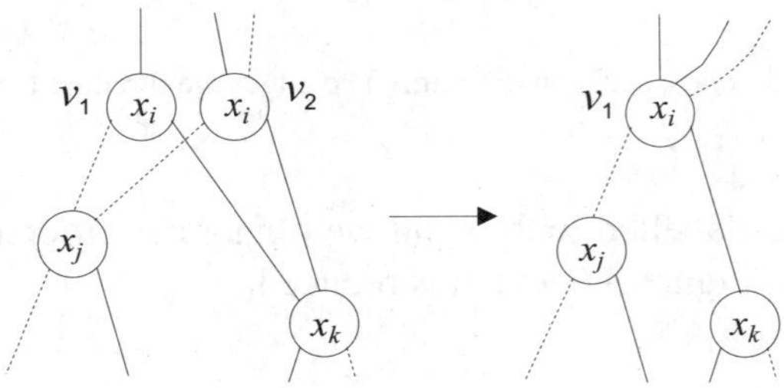

Figure 3.7b. Merging rule

canonical representation for each Boolean function and each variable ordering. Before proving this fact, let us introduce the notion of *isomorphic BDDs* which we are going to use in the proof of the canonicity of ROBDDs.

DEFINITION 3.10 (ISOMORPHIC BINARY DECISION DIAGRAMS)
Let $F_1 = (V_1, E_1, index_1, value_1)$ and $F_2 = (V_2, E_2, index_2, label_2)$ be two BDDs. F_1 and F_2 are said to be isomorphic, if there is a bijective function $\sigma : V_1 \rightarrow V_2$ such that for any node $v_1 \in V_1$ with $\sigma(v_1) = v_2$ either both v_1 and v_2 are terminal nodes with $value_1(v_1) = value_2(v_2)$ or both v_1 and v_2 are internal nodes with $index_1(v_1) = index(v_2)$, $\sigma(low(v_1)) = low(v_2)$, and $\sigma(high(v_1)) = high(v_2)$.

THEOREM 3.2 *Let $\pi \in \mathcal{P}er(\mathbb{N}_n)$ be any permutation of $\mathbb{N}_n$ and $<$ the variable order induced by π. Then, for each Boolean function $f \in \mathcal{B}_n$ there exists, up to isomorphy, only one ROBDD G over X_n of f such that G is ordered with respect to $<$.*

PROOF: We follow the proof given in [20] and prove the canonicity of ROBDDs by induction. First, we show that the ROBDD of a constant Boolean function is unique. Then, we assume that for each Boolean function which depends on at most $N \geq 0$ Boolean variables there exists a unique ROBDD.

Under this assumption, we prove that the ROBDD of a Boolean function which depends on $N + 1$ Boolean variables is unique, too.

Without loss of generality, we show that the ROBDD representation of the constant Boolean function $\mathbf{0} \in \mathcal{B}_n$ is unique. Let $G = (V, E, index, value)$ be a reduced ordered binary decision diagram of the Boolean function $\mathbf{0}$. G has a root node r_G and consists of $k \geq 0$ internal nodes. Each path starting from the root node r_G ends in a terminal node b with $value(b) = 0$ as $\phi(G) = \phi(r_G) = \mathbf{0}$. Thus, each terminal node of G has value 0 as label. As G is reduced, G contains only one terminal node. Now assume that k is greater than 0, i.e., G contains at least one internal node. As G is a finite graph, it also contains at least one internal node v whose low-edge and $high$-edge point to a terminal node. As there is only one terminal node both edges point to the same terminal node which is in contradiction to the condition that G is reduced. Thus, the ROBDD of the constant Boolean function $\mathbf{0}$ is unique. It consists of only one (terminal) node labelled with value 0. Analogously, we can prove that the ROBDD representation of the constant Boolean function $\mathbf{1} \in \mathcal{B}_n$ is unique.

Now, let $f \in \mathcal{B}_n$ be a Boolean function which depends on at least one Boolean variable. Thus, f is neither the constant Boolean function $\mathbf{0}$ nor the Boolean function $\mathbf{1}$. Let $G_1 = (V_1, E_1, index_1, value_1), G_2 = (V_2, E_2, index_2, value_2)$ be two ROBDDs of f which both are defined with respect to variable order $<$. As f is not constant, both G_1 and G_2 contain at least one internal node. Let r_{G_1} and r_{G_2} be the roots of G_1 and G_2, respectively. Let r_{G_1} be labelled with variable x_i, i.e., $index(r_{G_1}) = x_i$, and r_{G_2} with variable x_j. As

$$\phi(r_{G_1}) = f = \phi(r_{G_2}),$$

this results in

$$f = \left(x_i' \cdot \phi(low(r_{G_1}))\right) + \left(x_i \cdot \phi(high(r_{G_1}))\right)$$

and

$$f = \left(x_j' \cdot \phi(low(r_{G_2}))\right) + \left(x_j \cdot \phi(high(r_{G_2}))\right).$$

If x_i differs from x_j (without loss of generality, we assume $x_i < x_j$), then G_1 represents a Boolean function which depends on x_i and G_2 represents a Boolean function which does not depend on x_i. Thus, $index(r_{G_1}) = index(r_{G_2})$ must hold as both G_1 and G_2 represent the same Boolean function f. Thus, the equations

$$f = \left(x_i' \cdot \phi(low(r_{G_1}))\right) + \left(x_i \cdot \phi(high(r_{G_1}))\right)$$

$$f = \left(x_i' \cdot \phi(low(r_{G_2}))\right) + \left(x_i \cdot \phi(high(r_{G_2}))\right)$$

hold. By this, $\phi(low(r_{G_1})) = \phi(low(r_{G_2}))$ and $\phi(high(r_{G_1})) = \phi(high(r_{G_2}))$ follow.

As $\phi(low(r_{G_1}))$ is independent of variable x_i, the sub-ROBDDs rooted by $low(r_{G_1})$ and $low(r_{G_2})$ are isomorphic by induction, i.e., there exists a bijective function σ_{low} from the nodes of $low(r_{G_1})$ onto the nodes of $low(r_{G_2})$ such that for any node v_1 if $\sigma_{low}(v_1) = v_2$, then either both v_1 and v_2 are terminal nodes with $value_1(v_1) = value_2(v_2)$ or both v_1 and v_2 are nonterminal nodes with $index_1(v_1) = index_2(v_2)$, $\sigma_{low}(low(v_1)) = low(v_2)$, and $\sigma_{low}(high(v_1)) = high(v_2)$. As $\phi(high(r_{G_1})) = \phi(high(r_{G_2}))$ holds as well, there is such a bijective function σ_{high} from the nodes of $high(r_{G_1})$ onto the nodes of $high(r_{G_2})$, too.

Now, in order to prove that the ROBDDs G_1 and G_2 are isomorphic, we only have to prove that the function σ from the nodes of G_1 onto the nodes of G_2 defined by

$$\sigma(v) = \begin{cases} r_{G_2}, & \text{if } v = r_{G_1} \\ \sigma_{low}(v), & \text{if } v \text{ is a node of the ROBDD rooted by } low(r_{G_1}) \\ \sigma_{high}(v), & \text{if } v \text{ is a node of the ROBDD rooted by } high(r_{G_1}) \end{cases}$$

is well-defined and a bijective function.

Since for each node $v \in V_1$, which is contained in both the sub-ROBDD rooted by $low(r_{G_1})$ and the sub-ROBDD rooted by $high(r_{G_1})$, the equations $\phi(v) = \phi(\sigma_{low}(v))$ and $\phi(v) = \phi(\sigma_{high}(v))$ hold. Thus, $\sigma_{low}(v) = \sigma_{high}(v)$ has to hold as well as G_2 does not contain isomorphic sub-BDDs. This proves that function σ is well-defined.

Function σ is surjective because both σ_{low} and σ_{high} are surjective. It is injective because ROBDD G_1 does not contain isomorphic sub-BDDs. Thus, σ is bijective.

This completes the proof. ▪

Notation *As there is only one terminal node labelled with value* 0 *in ROBDDs, we simply call it terminal node* **0**. *Analogously, the terminal node labelled with value* 1 *is called terminal node* **1**.

In the restricted form defined in Definition 3.9 binary decision diagrams gained widespread application because of their canonicity and their efficiency in terms of time and space. Of course, Theorem 3.1 on page 25 also holds for ROBDDs. But many Boolean functions which occur in practical applications can efficiently be represented by ROBBDs.

3.2 Complexity of the essential operations

Now, let us investigate the complexities of the essential operations for ROBDDs of single-output Boolean functions of $\mathcal{B}_n$.

3.2.1 Boolean operations

All the unary and binary Boolean operations can efficiently be performed on ROBDDs. The common approach is to use the ternary if-then-else operator (ITE) defined by

$$ITE(f, g, h) = (f \cdot g) + (f' \cdot h)$$

which implements **if** f **then** g **else** h.
It is easy to see that

$$f \cdot g = ITE(f, g, \mathbf{0}),$$

$$f + g = ITE(f, \mathbf{1}, g),$$

and

$$f' = ITE(f, \mathbf{0}, \mathbf{1})$$

hold.
Thus, we have to present only the basic algorithm which implements the ITE-operator. The crucial idea of the algorithm is to use the Boole-Shannon decomposition of Boolean functions. By applying the Boole-Shannon decomposition it is easy to prove that

$$\begin{aligned} ITE(f, g, h) &= (x \cdot ITE(f_{x=1}, g_{x=1}, h_{x=1})) + \\ &\quad (x' \cdot ITE(f_{x=0}, g_{x=0}, h_{x=0})) \end{aligned}$$

holds. The terminal cases of this recursion are

$$\begin{aligned} ITE(\mathbf{1}, g, h) &= g, \\ ITE(\mathbf{0}, g, h) &= h, \\ ITE(f, \mathbf{1}, \mathbf{0}) &= f, \end{aligned}$$

and

$$ITE(f, g, g) = g.$$

This results in the algorithm shown in Figure 3.8.
The runtime of the algorithm is $O(|F| \cdot |G| \cdot |H|)$. To reach this runtime, the algorithm works with two internal tables:

- The *unique table* is a dictionary of the Boolean functions represented by the nodes which belong to a reduced ordered binary decision diagram. With the unique table the applications check whether a node labelled with variable x and children low and $high$ already exists. In the ITE-algorithm, for example, the unique table is used by function FIND_OR_ADD_UNIQUE_TABLE. If

```
ITE(F,G,H)
begin

  if TERMINAL_CASE("ITE",F,G,H,result)
  // true iff the recursion can be aborted because of a special case
  then return result;
  fi;

  if FIND_IN_COMPUTED_TABLE("ITE",F,G,H,result)
  // true iff ITE(F,G,H) has already been computed
  then return result;
  fi;

  x = smallest variable occurring in F, G, and H;
  high = ITE(F_{x=1}, G_{x=1}, H_{x=1});
  low = ITE(F_{x=0}, G_{x=0}, H_{x=0});

  if high==low
  then result = high;
  else result = FIND_OR_ADD_UNIQUE_TABLE(x,low,high);
       // returns the ROBDD of (x' · low) + (x · high)
  fi;

  INSERT_IN_COMPUTED_TABLE("ITE",F,G,H,result);
  return result;

end;
```

Figure 3.8. The ITE-algorithm. Note that it is necessary that the variables of the three ROBDDs F, G, and H have to be ordered with respect to the same variable order.

the function finds an existing node labelled with x which has low and $high$ as 0-successor and 1-successor, respectively, it outputs the address of this node which is interpreted as root node of the ROBDD just synthesized. Otherwise, a new node is created and added to the unique table. With this, the merging rule is implicitly applied already during the synthesis algorithm. The deletion rule is applied by the lower **if**-statement of the algorithm. Thus the BDD returned by the ITE-algorithm is always a reduced ordered BDD.

- The *computed table* is used to store already computed results of previous calls of the ITE-algorithm. Thus, the ITE-algorithm is called at most once for each combination of nodes in F, G, and H which results in the overall time complexity of $O(|F| \cdot |G| \cdot |H|)$ [43].

Both tables are best implemented by hash tables. Since it is usual when analyzing the computational complexity of decision diagram algorithms we assume "ideal" hash tables for the unique table and the computed table, i.e., look-ups and insertions can be performed in constant time. (For more details, we refer to [56].) Thus, the binary operators can be efficiently executed if ROBDDs are used.

THEOREM 3.3 *Let F and G be two ROBDDs of Boolean functions of $\mathcal{B}_n$ which are ordered with respect to the same variable order. Then, the ROBDD of $\phi(F) \cdot \phi(G)$ and the ROBDD of $\phi(F) + \phi(G)$ can be computed in time $O(|F| \cdot |G|)$. The ROBDD of $\phi(F)'$ can be computed in time $O(|F|)$.*

3.2.2 Cofactor computation

Let us consider an ROBDD F over X_n of a Boolean function $f \in \mathcal{B}_n$ and any variable $x_i \in X_n$. We have to redirect all edges leading to nodes labelled with x_i to the c-successor of these nodes in order to compute the cofactor $f_{x_i=c}$ [47]. As, the resulting ordered binary decision diagram is not necessarily reduced, a reduction algorithm has to be applied to the result. Both steps can be performed in time $\mathcal{O}(|F|)$. Before the transformation, a copy of ROBDD F has to be created in order to conserve the representation of f.

Cofactor computation is rather trivial if x_i is the first variable of the variable order $<_F$ of ROBDD F, i.e., $x_i = x_{\pi_F(1)}$. Let r_F be the root node of F, then either r_F is not labelled with x_i or it is labelled with x_i. In the first case, f is independent of x_i, i.e., $f = f_{x_i=1} = f_{x_i=0}$. In the second case, $low(r_F)$ and $high(r_F)$ are the ROBDDs of the negative cofactor $f_{x_i=0}$ and positive cofactor $f_{x_i=1}$, respectively. Thus, in this special case the cofactor computation has constant runtime. It has to return the address of $low(r_F)$ and $high(r_F)$, only.

3.2.3 Substitution by functions

Given the ROBDDs F and G of two Boolean functions f and g of $\mathcal{B}_n$ defined over the same set X_n of variables, the computation of $f_{x_i=g}$ can be solved easily by using the if-then-else operator ITE. Remember that f can be decomposed as

$$f = (x_i' \cdot f_{x_i=0}) + (x_i \cdot f_{x_i=1}).$$

Both cofactors $f_{x_i=0}$ and $f_{x_i=1}$ are independent of x_i. Thus

$$(f_{x_i=g}) = (g' \cdot f_{x_i=0}) + (g \cdot f_{x_i=1})$$

holds. The corresponding ROBDD is computed by

$$ITE(G, F_{x_i=1}, F_{x_i=0})$$

where $F_{x_i=1}$ and $F_{x_i=0}$ denote the ROBDDs of the positive and negative cofactor of f with respect to x_i, respectively. The time complexity of this operation is $\mathcal{O}(|G| \cdot |F|^2)$.

3.2.4 Satisfiability and tautology check

Both, the satisfiability check and tautology check applied to a ROBDD F of a Boolean function of $\mathcal{B}_n$ can be performed in constant time. Just check if the ROBDD F consists of terminal node **0** or terminal node **1** only.

3.2.5 Satisfy count computation

Let F be the ROBDD over X_n of a Boolean function $f \in \mathcal{B}_n$ and r_F the root node of F, the satisfy count $|f|$ of f can be computed by a bottom-up algorithm which computes the number of paths $|w|$ from each node w to the terminal node l_1 which is labelled with 1. Actually, it computes the number of paths from each node w to terminal node l_1 in the *complete* BDD which is obtained when applying in the opposite direction the deletion rule to the ROBDD F such that each variable is encountered exactly once on each path from root node r_F to a terminal node.

There is no path from the terminal node l_0 labelled with 0 to l_1 and there is exactly one path from l_1 to l_1. Therefore, $|l_0|$ and $|l_1|$ are 0 and 1, respectively. For each internal node w of F with successors w_0 and w_1, the number of paths $|w|$ from w to l_1 is given by

$$|w| = |w_0| \cdot 2^{level_F(w_0) - level_F(w) - 1} + |w_1| \cdot 2^{level_F(w_1) - level_F(w) - 1}.$$

Then, the satisfy count $|f|$ of f is given by

$$|f| = |r_F| \cdot 2^{level_F(r_F) - 1}.$$

The complexity of this algorithm in terms of time is $\mathcal{O}(|F|)$. It is summarized in Figure 3.9.

3.2.6 ROBDD minimization

The efficiency of the algorithms just considered strongly depends on the size of the ROBDD representations of the Boolean functions under consideration — the larger the ROBDD representations the more costly the procedures. In accordance to this fact, many researchers have developed algorithms and heuristics for ROBDD minimization [26, 41, 44, 50, 51, 52, 66, 68, 93, 102, 103, 114, 115, 125, 128, 132].

Note that ROBDDs are a canonical representation of Boolean functions for a fixed variable order only. If the variable order is changed, then the ROBDDs change in general as well. In many cases, the choice of the variable order is crucial as the size of the ROBDDs can strongly depend on the variable order

```
SATISFY_COUNT(F)
begin
  |l_1|=0; |l_0|=1;

  for j from n downto 1
  do
    forall node w of the jth level
    do
    |w| = |low(w)|·2^(level(low(w))−level(w)−1) + |high(w)|·2^(level(high(w))−level(w)−1);
    od;
  od;

  return |r| · 2^(level(r_F)−1);

end;
```

Figure 3.9. A linear bottom-up algorithm for satisfy count computation.

chosen. Figure 3.10, which is taken from [20], illustrates this remark. The Boolean function which is defined by the Boolean formula

$$(x_1 \cdot x_2) + (x_3 \cdot x_4) + (x_5 \cdot x_6)$$

is represented twice, once by an ROBDD with respect to the variable order

$$x_1 < x_2 < x_3 < x_4 < x_5 < x_6$$

and once by an ROBDD with respect to the variable order

$$x_1 < x_3 < x_5 < x_2 < x_4 < x_6.$$

The ROBDDs consist of 6 and 14 internal nodes, respectively. It is straightforward to generalize this example. For a fixed $n \in \mathbb{N}$, consider the Boolean function $f^{(n)} \in \mathcal{B}_{2n}$ defined by the Boolean formula

$$(x_1 \cdot x_2) + \ldots + (x_{2i-1} \cdot x_{2i}) + \ldots + (x_{2n-1} \cdot x_{2n}).$$

The ROBDD of $f^{(n)}$ with respect to variable order

$$x_1 < x_2 < \ldots < x_{2n-1} < x_{2n}$$

consists of $2n$ internal nodes, whereas the ROBDD of $f^{(n)}$ with respect to variable order

$$x_1 < x_3 < \ldots < x_{2n-3} < x_{2n-1} < x_2 < x_4 < \ldots < x_{2n-1} < x_{2n}$$

consists of at least 2^n nodes. The latter statement is followed by the fact that there must be at least as many nodes as there are different assignments to the variables $x_1, x_3, \ldots, x_{2i-1}, \ldots, x_{2n-1}$. More examples where the choice of the variable order determines whether the number of nodes is linear or exponential are presented in [147].

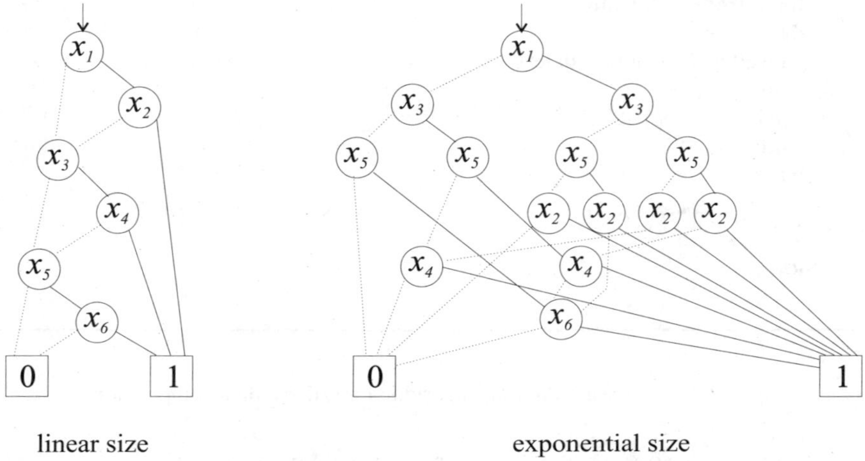

Figure 3.10. ROBDDs of $(x_1 \cdot x_2) + (x_3 \cdot x_4) + (x_5 \cdot x_6)$ wrt. different variable orders.

Unfortunately, the ROBDD minimization problem is NP-hard. Bollig, Savicky, and Wegener [11, 12] showed that the problem

> **Given** an ROBDD of a Boolean function $f \in \mathcal{B}_n$,
>
> **compute** a minimum-sized ROBDD of f, i.e., compute a variable order which induces a ROBDD of f with a minimum number of internal nodes

is NP-hard. Meinel [101] proved that the decision problem

> **Given** a Boolean Network $\mathcal{N}$ of a Boolean function $f \in \mathcal{B}_n$ and a permutation $\pi \in \mathcal{P}er(\mathbb{N}_n)$,
>
> **decide** whether the size of the ROBDD of f induced by π is minimum

is NP-hard as well.

Thus, for both situations, the situation that the Boolean function is represented by a circuit and the situation that it is already represented by an ROBDD, the optimization problem of finding an optimal variable order can only be handled by heuristics, in general.

Since the subject of ROBDD minimization has been discussed at full length in many publications, already, we do not address the issue of finding optimal variable orders in this book. We refer to the surveys given in [42, 43, 47] and briefly present the *sifting* algorithm introduced by Rudell in 1993 [125] only. Most of the heuristics presented in literature are based on Rudell's idea.

Given an ROBDD of a Boolean function, the sifting algorithm selects a variable x_i and searches for the position of this variable which minimizes the ROBDD size. The positions of the rest of the variables do not change. This process is iterated until every variable is sifted once. The local minimization step is realized by exchanging neighboring variables, i.e., variable x_i is first pushed to the end of the variable order by exchanging it with the following variables with respect to the variable order, step by step. Then, it is pushed to the first position of the variable order. Finally, it is moved to its (locally) optimal position.

The basic exchange operation is a local operation. The goal of the operation applied to the variables $x_{\pi_F(i)}$ and $x_{\pi_F(i+1)}$ of an ROBDD F is to compute the ROBDD G of $\phi(F)$ with

$$\pi_G(j) = \begin{cases} \pi_F(j), & \text{if } j \notin \{i, i+1\} \\ \pi_F(i), & \text{if } j = i+1 \\ \pi_F(i+1), & \text{if } j = i. \end{cases}$$

Now, look at a node w in F which is labelled with $x_{\pi_F(i)}$ and which has at least one successor labelled with $x_{\pi_F(i+1)}$. It follows directly from the Boole-Shannon decomposition that the exchange operation is correctly performed by redirecting the corresponding low- and high-edges as shown in Figures 3.11a-3.11c. Thus, the ROBDD of F remains unchanged above variable $x_{\pi_F(i)}$ and below variable $x_{\pi_F(i+1)}$. For the ith and $(i+1)$th level of G, it can be proven that

$$1/2 \cdot |level_F^{-1}(i+1)| \leq |level_G^{-1}(i)| \leq |level_F^{-1}(i)| + |level_F^{-1}(i+1)|$$

and

$$1 \leq |level_G^{-1}(i+1)| \leq 2 \cdot |level_F^{-1}(i)|$$

hold [44] as $\phi(F)$ depends on both variables $x_{\pi_F(i)}$ and $x_{\pi_F(i+1)}$. If the nodes labelled with $x_{\pi_F(i)}$ can be traversed in linear time, this algorithm is very efficient. The runtime is linear with respect to the number of nodes labelled with $x_{\pi_F(i)}$.

3.3 Generalizations of Binary Decision Diagrams

There is quite a large number of extensions of the basic ROBDD concept to represent Boolean functions of $\mathcal{B}_n$ which have been proposed in literature. These investigations were motivated by the fact that, although ROBDDs are more effective than other representations of Boolean functions, *there are* Boolean functions occuring in practical applications for which ROBDDs are too large to

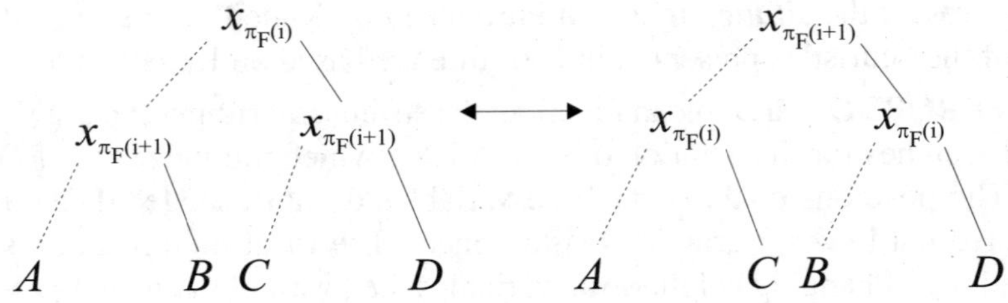

Figure 3.11a. Exchanging of neighboring variables: first case. A, B, C, and D denote sub–BDDs. The circles of the internal nodes have been omitted for a better graphical representation.

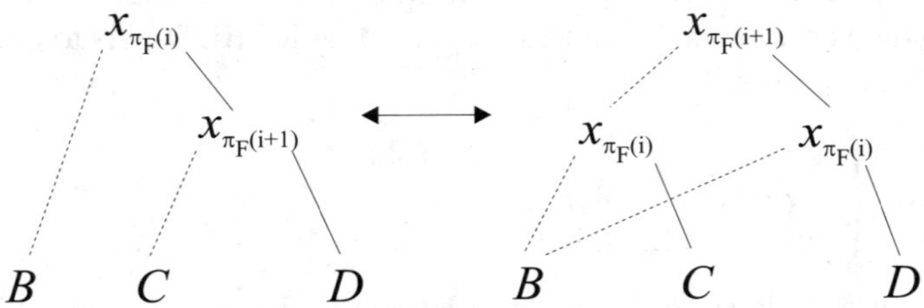

Figure 3.11b. Exchanging of neighboring variables: second case

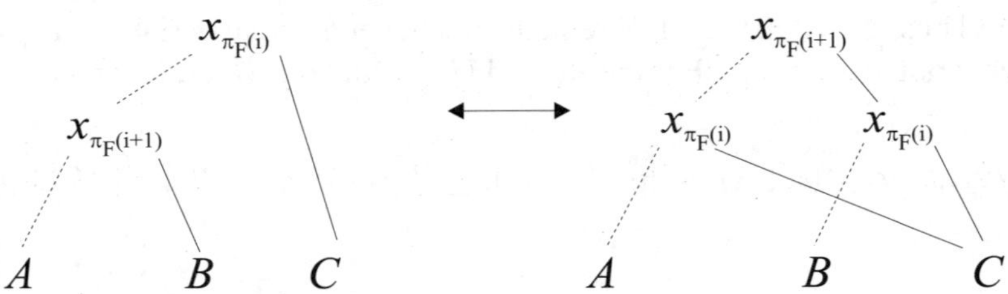

Figure 3.11c. Exchanging of neighboring variables: third case

be stored in computer memory. For example, it is well-known that ROBDDs of the multiplication of two n-bit integers have at least $2^{n/8}$ nodes [20]. In this section, we briefly deal with two of these extensions, *ROBDDs with complemented edges* and decision diagrams based on different decomposition rules, namely *functional* and *Kronecker functional decision diagrams*. Further variations of decision diagrams for Boolean functions can be found in [21, 126]. Section 4 goes into a further generalization of BDDs, *binary moment diagrams*, which is a representation of pseudo–Boolean functions.

3.3.1 ROBDDs with complemented edges

In some implementations a slightly different version of ROBDDs, namely ROBDDs with complemented edges, is used. Complemented edges were suggested by [3, 15, 104]. On each edge (and each pointer to a root node) there may be a complement attribute. Such a *complemented edge* indicates that the interpretation of the node the edge points at is complemented when the node is reached via this edge.

ROBDDs with complement edges where complement attributes only occur on high-edges are a canonical representation of Boolean functions, too. Note, that the reduction rules have to be changed slightly in a straightforward manner and complement attributes on low-edges can be easily removed by a bottom-up pass through the ROBDD.

The number of nodes of ROBDDs with complemented edges is usually smaller than the number of nodes in ROBDDs without complemented edges – the size of ROBDDs with complemented edges can be half the size of the corresponding ROBDD without complemented edges. The asymptotical complexity of the essential operations remains unaffected. However the description of the algorithms are more complicated for ROBDDs with complemented edges. In order to avoid to become entangled in technical details, we use ROBDDs *without* complemented edges throughout this book.

3.3.2 Functional Decision Diagrams

In [72, 73], *ordered functional decision diagrams* (FDD) have been proposed as representation of Boolean functions. They are syntactically defined as ordered BDDs but the Shannon decomposition is replaced by the *positive Davio decomposition*

$$f = f_{x_i=0} \oplus (x_i \cdot \partial f / \partial x_i).$$

More formally, let v be an internal node labelled with x_i, then the Boolean function

$$\phi(v) = \phi(low(v)) \oplus (x_i \cdot \phi(high(v)))$$

is represented at v. The Boolean function represented by an ordered FDD can be described using activated paths [58], i.e., the Boolean function evaluates to 1 with respect to an assignment $\alpha \in \{0,1\}^n$ if and only if there is an odd number of activated paths from the root node to the terminal node **1** with respect to α.

DEFINITION 3.11 (ACTIVATED PATHS IN FDDS) *Let G be an FDD over X_n. Let v be an internal node of G labelled with x_i and $e := (v, w)$ be an* *outgoing edge of* v*. Edge* e *is called activated with respect to an assignment* $\alpha \in \{0,1\}^n$ *if either* $w = low(v)$ *or* $w = high(v)$ *and* $\alpha_i = 1$*. This can be notified by the activation condition* $activate(e)$ *of* e *which is defined by*

$$activate(e) = \begin{cases} 1 & , \text{if } w = low(v) \\ x_i & , \text{if } w = high(v). \end{cases}$$

A path $p = (e_1, e_2, \ldots, e_h)$ *is called activated with respect to* α*, if every edge of* p *is activated with respect to* α *, i.e., the activation condition is described by*

$$activate(p) = \prod_{j=1}^{h} activate(e_j).$$

Symbol $\prod$ *denotes the iterated logical-and operator.*

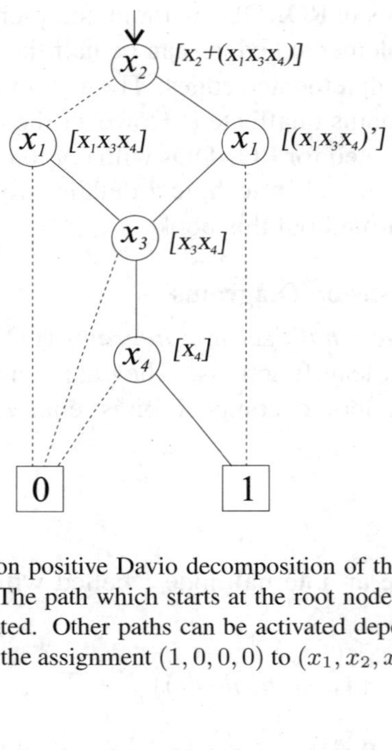

Figure 3.12. FDD based on positive Davio decomposition of the Boolean function $f \in \mathcal{B}_4$ given by $(x_1x_3x_4) + x_2$. The path which starts at the root node and which consists of low-edges only is always activated. Other paths can be activated dependent on the assignment to the variables. For example, the assignment $(1,0,0,0)$ to (x_1,x_2,x_3,x_4) induces two activated paths in total.

For illustration we refer to Figure 3.12 which shows an FDD for the Boolean function $f \in \mathcal{B}_4$ given by $(x_1 \cdot x_3 \cdot x_4) + x_2$. Assuming that the low-successor and the high-successor of the root represent the Boolean functions given by the Boolean formulae $x_1 \cdot x_3 \cdot x_4$ and $(x_1 \cdot x_3 \cdot x_4)'$, respectively, the root represents the Boolean function given by

$$(x_1 \cdot x_3 \cdot x_4) \oplus (x_2 \cdot (x_1 \cdot x_3 \cdot x_4)') = (x_1 \cdot x_3 \cdot x_4) + x_2$$

as for all $a, b \in \{0, 1\}$ the equation $a \oplus (b \cdot a') = a + b$ holds.

Besides the Boole-Shannon expansion and the positive Davio decomposition, there is a third decomposition, the negative Davio decomposition

$$f = f_{x_i=1} \oplus (x_i{}' \cdot \partial f / \partial x_i).$$

Of course, there is also a variant of ordered FDDs based on the *negative Davio decomposition* where the Boolean function $\phi(v)$ represented by an internal node v is given by

$$\phi(v) = \phi(low(v)) \oplus (x_i{}' \cdot \phi(high(v))).$$

In the previous section we have introduced two reduction rules to be able to remove redundancy in ROBDDs. While the *merging rule* for ordered BDDs can be applied to ordered FDDs as well, we need different *deletion rules* for ordered Davio FDDs. Here, an internal node v is superfluous, if the high-edge of v points to the terminal node 0. This is because of

$$\begin{aligned} \phi(v) &= \phi(low(v)) \oplus (x_i^\epsilon \cdot \phi(high(v))) \\ &= \phi(low(v)) \oplus (x_i^\epsilon \cdot 0) \\ &= \phi(low(v)) \oplus 0) \\ &= \phi(low(v)) \end{aligned}$$

for $\epsilon \in \{0, 1\}$. Reduced ordered FDDs are a canonical representation of Boolean functions.

In [40, 46], ordered BDDs, ordered positive Davio FDDs, and ordered negative Davio FDDs are combined to a new variant called ordered *Kronecker functional decision diagrams* (KFDD). The main idea is that at different internal nodes different decompositions may be chosen with the constraint that the decompositions of internal nodes labelled with the same variable are the same. That is, the decomposition types of the internal nodes can be described by a mapping $d : X_n \rightarrow \{S, pD, nD\}$. S, pD, and pD denote the Boole-Shannon decomposition, the positive Davio decomposition, and the negative Davio decomposition, respectively. The Boolean function represented by an ordered KFDD can be described using activated paths as well [58].

DEFINITION 3.12 (ACTIVATED PATHS IN KFDDS) *Let G be a KFDD over X_n, v an internal node of G labelled with x_i and $e := (v, w)$ an outgoing edge of v. The activation condition $activate(e)$ of edge e can be described as follows:*

$$activate(e) = \begin{cases} 1 & , \textit{ if } w = low(v) \textit{ and } d(x_i) \in \{pD, nD\} \\ x_i{}' & , \textit{ if either } w = low(v) \textit{ and } d(x_i) = S \\ & \quad \textit{ or } w = high(v) \textit{ and } d(x_i) = nD. \\ x_i & , \textit{ if } w = high(v) \textit{ and } d(x_i) \in \{S, pD\}. \end{cases}$$

As above, the activation condition of a path $p = (e_1, e_2, \ldots, e_h)$ *is described by*

$$activate(p) = \prod_{j=1}^{h} activate(e_j).$$

THEOREM 3.4 ([58]) *Let G be a KFDD of a Boolean function* $f \in \mathcal{B}_n$ *and P the set of the paths from the root node of G to the terminal node* **1**. *Then*

$$\bigoplus_{p \in P} activate(p)$$

is a description of f.

Reduced ordered KFDDs are a canonical representation of Boolean functions as well. Of course, as ROBDDs and reduced ordered Davio FDDs are reduced ordered KFDDs, the size of the minimum-sized reduced ordered KFDD of a Boolean function f is not larger than the minimum of the sizes of ordered BDDs and ordered Davio FDDs of f. However, the problem with reduced ordered FDDs and KFDDs is that the synthesis operations $\cdot$ and $+$ can exponentially blow up the size of the decision diagrams [6, 40] while the synthesis operations for ROBDDs can be performed by polynomial time algorithms.

In [5], Becker et.al. showed that when considering canonical representations with complemented edges there is no further decomposition type besides Boole-Shannon, positive Davio, and negative Davio that can reduce the size of ordered decision diagrams.

4. Representations of pseudo–Boolean functions

In this section, we introduce how to represent pseudo–Boolean functions by multiplicative binary moment diagrams (*BMDs). *BMDs have been introduced by Bryant and Chen [22, 23]. They incorporate two novel features compared to ROBDDs. They are based on a decomposition of a linear function in terms to its moments. Furthermore, they have weights associated with their edges which are combined multiplicatively. These features are derived from the Davio decomposition used by FDDs and the additive edge weights used in so-called *Edge-Valued BDDs* [84].

*BMDs have been motivated by the necessity to represent pseudo–Boolean functions $f \in \mathcal{Z}_n$. As already shown in Theorem 2.3, such functions can be decomposed by *moment decomposition* with respect to a variable x_i which is defined as

$$f = f_{x_i=0} + x_i \cdot f_{\dot{x}_i},$$

where $\cdot$ and $+$ denote integer multiplication[1] and integer addition[2], respectively. Note, that x_i is interpreted here as an integer-valued function, having the image of zero and one. The negative cofactor $f_{x_i=0}$ is called *constant moment* as it is added independently of the assignment of x_i. $f_{\dot{x}_i}$ which is given by $\partial f/\partial x_i$ is called *linear moment* of f with respect to variable x_i.

4.1 Basic definitions

We start with the definition of *binary moment diagrams* before descending to *multiplicative* binary decision diagrams. The structure of ordered binary moment diagrams is similar to that of ordered BDDs, except that the terminal nodes are labelled with integer values and they are based on moment decomposition.

DEFINITION 3.13 (BINARY MOMENT DIAGRAM) *A binary moment diagram over X_n is a rooted, directed, acyclic graph $G = (V, E, index, value)$. An internal node $v \in V$ has as label a variable $index(v) \in X_n$ and exactly two successors $low(v)$ and $high(v)$. The edges pointing to the successors are named low-edge, or 0-edge, and high-edge, or 1-edge, respectively. The successors themselves are named low-successor, or 0-successor, and high-successor, or 1-successor, respectively. The terminal nodes have no successor and are labelled with integer values.*

A binary moment diagram is called ordered if the variables appear in the same order on every path from the root to the terminal nodes.

DEFINITION 3.14 (INTERPRETATION OF BINARY MOMENT DIAGRAMS) *Let $G = (V, G, index, value)$ be a binary moment diagram over X_n and $v \in V$ be a node of G. The pseudo–Boolean function $\phi(v) \in \mathcal{Z}_n$ which is computed at node v is defined as follows.*

- *If v is a terminal node, then $\phi(f)$ is the constant pseudo–Boolean function with $\phi(f)(\alpha) = value(v)$ for each $\alpha \in \{0,1\}^n$.*

- *If v is an internal node and $index(v) = x_i$ then*

$$\phi(v) = \phi(low(v)) + x_i \cdot \phi(high(v))$$

where $\cdot$ and + denote integer multiplication and integer addition, respectively.

Once again, the concept of activated paths can be used in order to evaluate a binary moment diagram. We can apply the same definition as for FDDs (see Definition 3.11). The evaluation process adds the values of the terminal nodes of the activated paths which start at the root node and end at terminal nodes.

THEOREM 3.5 *Let G be a binary moment diagram of a pseudo–Boolean function $f \in \mathcal{Z}_n$ and P the set of the paths which start at the root node r_G of G and end at terminal nodes. Then*

$$\sum_{p \in P} activate(p) \cdot value(p)$$

is a description of f. $value(p)$ denotes the value of the terminal node of path p.

In the following, we consider reduced ordered binary moment diagrams only. We call them BMDs, as they are a canonical representation of pseudo–Boolean functions.

DEFINITION 3.15 (REDUCED ORDERED BINARY MOMENT DIAGRAM) *An ordered binary moment diagram is called reduced if there are no isomorphic subgraphs and unnecessary nodes in the binary moment diagram, i.e., there are no two distinct nodes that are root nodes of isomorphic subgraphs and there is no internal node with the high-edge pointing to a terminal node labelled with value 0.*

BMD is going to be used as short hand for reduced ordered binary moment diagram, in the following.

Multiplicative binary moment diagrams (*BMD) have the same structure as BMDs, except that a *BMD makes use of a common integer factor in the constant and linear moment. It extracts this factor and places it as a so-called *edge-weight* on the incoming edge to the node. In this context, we interpret the pointer to the root node of a *BMD as an edge. Let us denote the weight of an edge e by ω_e. Then, a *BMD F with root node r_F labelled with x_i and weight ω_F on the incoming edge of r_F represents the pseudo–Boolean function $\phi(\omega_F, r_F)$ defined by

$$\phi(\omega_F, r_F) = \omega_F \cdot (\phi(\omega_{(r_F, low(r_F))}, low(r_F)) + x_i \cdot \phi(\omega_{(r_F, high(r_F))}, high(r_F))).$$

If r_F is a terminal node, then $\phi(\omega_F, r_F)$ is the constant function defined by

$$\phi(\omega_F, r_F) = \omega_F \cdot value(r_F).$$

We denote $\phi(\omega_F, r_F)$ by $\phi(F)$, in the following.

To make the *BMD-representation canonical, some further restrictions are necessary. These restrictions are called *normalizing rules*.

1. A *BMD may contain at most two terminal nodes which are labelled with value 0 and value 1, respectively. We call them terminal node **0** and terminal node **1**, respectively.

2. The weights of all the edges must be different from 0.

3. If an edge points to the terminal node **0**, its weight must be 1. If a low-edge points to terminal node **0**, the weight of the corresponding high-edge has to be 1, too.

4. The weights of all the low-edges must be nonnegative values.

5. The weights of the outgoing edges of an internal node must have greatest common divisor 1, i.e.,

$$\gcd(\omega_{(v,low(v))}, \omega_{(v,high(v))}) = 1$$

must hold for each internal node v.

6. A *BMD has to be reduced, i.e., there must not be two distinct nodes that are root nodes of isomorphic subgraphs and there must not be an internal node with the high-edge pointing to terminal node **0**.

For illustration, consider Figure 3.13. Edge-weights are written in square boxes on the corresponding edges. Edges without square boxes have edge-weight 1. For a better understanding, we have inserted the pseudo–Boolean functions represented at the nodes of the *BMD into the figure.

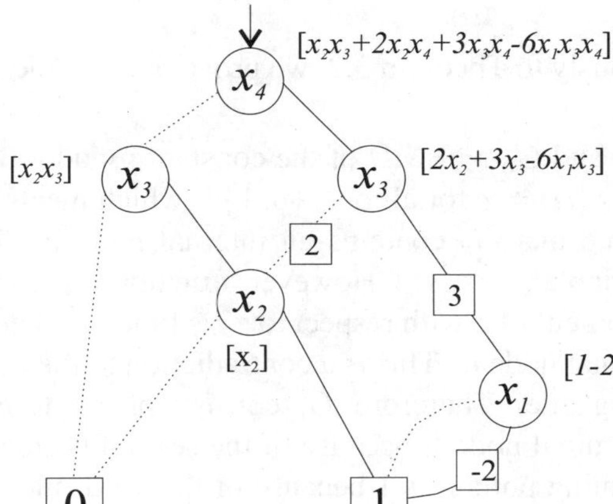

Figure 3.13. *BMD of the function $x_2 \cdot x_3 + 2 \cdot x_2 \cdot x_4 + 3 \cdot x_3 \cdot x_4 - 6 \cdot x_1 \cdot x_3 \cdot x_4$

Now, before proving canonicity, let us show the following lemma which we are going to apply throughout the proof of Theorem 3.6.

LEMMA 3.1 *Let f be a pseudo–Boolean function with $f \neq \mathbf{0}$, F be a normalized *BMD of f with root note r_F, and $d \geq 2$ a natural number. If d is a* *divisor of f, i.e., $f = d \cdot g$ for some adequate pseudo–Boolean function g, then the weight ω_F of the incoming edge of r_F is a multiple of d.*

PROOF: Let us prove the lemma by induction.

First, we assume that r_F is a terminal node, i.e., f is a constant pseudo–Boolean function $\boldsymbol{c} \in \mathcal{Z}_n$ defined by $\boldsymbol{c}(\alpha) = c$ for all $\alpha \in \{0,1\}^n$, $c \in \mathbb{Z}$. Because of the first normalizing rule, the weight ω_F is c which proves the statement for this case.

Now, we assume that both successors $low(r_F)$ and $high(r_F)$ are terminal nodes, i.e., $f = \omega_F \cdot (\omega_{(r_F,low(r_F))} + \omega_{(r_F,high(r_F))} \cdot x_i)$ for $x_i = index(r_F)$. If d is not a divisor of ω_F, the greatest common divisor of $\omega_{(r_F,low(r_F))}$ and $\omega_{(r_F,high(r_F))}$ is greater than 1, which is in contraction to the fifth normalizing rule.

Now, let us consider the general case. Assume that d is not a divisor of ω_F, then there is a constant $\delta \geq 2$ which is a divisor of both $\phi(\omega_{(r_F,low(r_F))}, low(r_F))$ and $\phi(\omega_{(r_F,high(r_F))}, high(r_F))$. By induction, it follows that δ is a divisor of both edge weights $\omega_{(r_F,low(r_F))}$ and $\omega_{(r_F,high(r_F))}$. This contradicts the fact that F meets the fifth normalizing rule as $\delta \geq 2$. ∎

THEOREM 3.6 *Let $\pi \in \mathcal{P}er(\mathbb{N}_n)$ be any permutation of $\mathbb{N}_n$. Then for each pseudo–Boolean function $f \in \mathcal{Z}_n$ there exists, up to isomorphy, only one *BMD F over X_n which is ordered with respect to π and which meets the normalizing rules.*

PROOF: Analogously to Theorem 3.2, we prove the canonicity of *BMDs by induction.

Let c be an integer and G_c a *BMD of the constant pseudo–Boolean function $\boldsymbol{c} \in \mathcal{Z}_n$ defined by $\boldsymbol{c}(\alpha) = c$ for all $\alpha \in \{0,1\}^n$, which meets the normalizing rules. Now, assume that G_c contains an internal node v. Then, node v is labelled with a variable $x_i \in X_n$. However, function $\boldsymbol{c}$ is independent of x_i. Thus, the linear moment of $\boldsymbol{c}$ with respect to x_i is function $\mathbf{0}$ and the high-edge of v points to terminal node $\mathbf{0}$. This is a contradiction to the fact that G_c meets the last normalizing rule. Therefore, G_c consists of one terminal node t. If $c = 0$, then t is terminal node $\mathbf{0}$ because of the second normalizing rule. The weight of the incoming pointer is 1 because of the third rule. If $c \neq 0$, then t is terminal node $\mathbf{1}$ and the weight of the incoming pointer has to be c because of the first normalizing rule. This proves that there is only one *BMD of the constant pseudo–Boolean function $\boldsymbol{c}$.

Now, let $f \in \mathcal{Z}_n$ be a pseudo–Boolean function which depends on at least one Boolean variable. Let G and H be two normalized *BMDs of f with respect to the variable order induced by π. Furthermore, let r_G and r_H be the root nodes of G and H, respectively. As G and H represent the same pseudo–Boolean function f and neither the high-edge of r_G nor the high-edge of r_H

points to terminal node **0**, f depends on both variable $index(r_G)$ and variable $index(r_H)$. It follows that $index(r_G) = index(r_H)$ holds because G and H are defined with respect to the same variable order. As the constant moment and the linear moment of a pseudo–Boolean function with respect to a variable is uniquely defined, the equations

$$\phi(\omega_G \cdot \omega_{(r_G, low(r_G))}, low(r_G)) = \phi(\omega_H \cdot \omega_{(r_H, low(r_H))}, low(r_H))$$

and

$$\phi(\omega_G \cdot \omega_{(r_G, high(r_G))}, high(r_G)) = \phi(\omega_H \cdot \omega_{(r_H, high(r_H))}, high(r_H))$$

hold, too.

By induction, there is, up to isomorphy, only one *BMD of

$$\phi(\omega_G \cdot \omega_{(r_G, low(r_G))}, low(r_G))$$

which is defined with respect to π and meets the normalizing rules. Analogously, there is, up to isomorphy, exactly one such *BMD of

$$\phi(\omega_G \cdot \omega_{(r_G, high(r_G))}, high(r_G))$$

as well.

Because of Lemma 3.1 and the fact that both *BMD G and *BMD H meet the normalizing rules, the two equations

$$\omega_G \cdot \omega_{(r_G, low(r_G))} = \omega_H \cdot \omega_{(r_H, low(r_H))}$$
$$\omega_G \cdot \omega_{(r_G, high(r_G))} = \omega_H \cdot \omega_{(r_H, high(r_H))}$$

hold.

Because of the fourth reduction rule, both weights $\omega_{(r_G, low(r_G))}$ and $\omega_{(r_H, low(r_H))}$ are nonnegative. Now assume, that the weights $\omega_{(r_G, low(r_G))}$ and $\omega_{(r_H, low(r_H))}$ differ. It follows that ω_G and ω_H also differ because of the two equations above. However, this is in contradiction to the fact that the constant and linear moment of f with respect to a variable is well-defined and Lemma 3.1. Thus, the equations

$$\omega_G = \omega_H,$$
$$\omega_{(r_G, low(r_G))} = \omega_{(r_H, low(r_H))},$$
$$\omega_{(r_G, high(r_G))} = \omega_{(r_H, high(r_H))}$$

hold and the theorem follows analogously to the proof of Theorem 3.2. ■

For the sake of clarity, the graphical representations of *BMDs shown in the following illustrations have more than two terminal nodes, in general. However, it is straightforward to transform them into representations which meet the normalizing rules (see Section 4.2).

To demonstrate the difference between BMDs and *BMDs, we refer to Figure 3.14a and Figure 3.14b which show the BMD and the *BMD of the 3-bit integer multiplier, i.e., the pseudo–Boolean function $(\sum_{i=0}^{2} x_i \cdot 2^i) \cdot (\sum_{i=0}^{2} y_i \cdot 2^i)$.

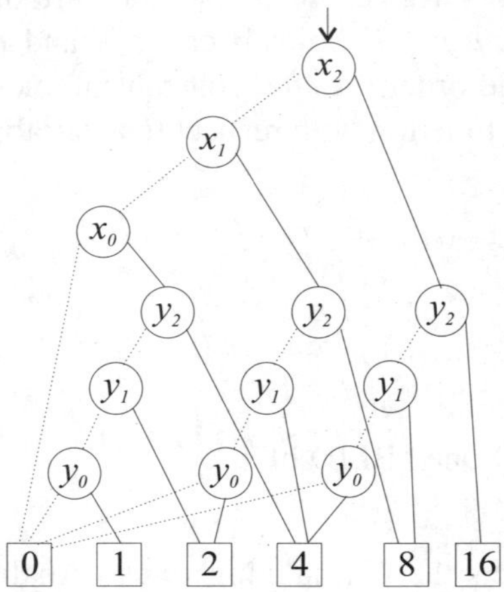

Figure 3.14a. BMD of the 3-bit integer multiplier $x \cdot y$

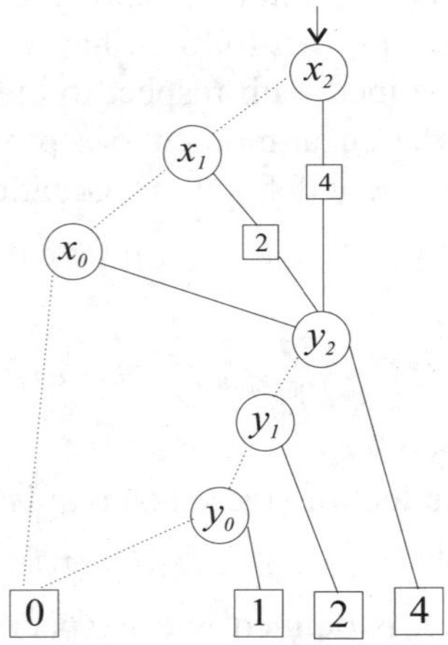

Figure 3.14b. *BMD of the 3-bit integer multiplier $x \cdot y$

Note that the representational power of *BMDs is greater than that of BMDs [4, 22]. The minimum sized *BMD of a pseudo–Boolean function f never contains more nodes than a BMD representation of f. Additionally, there is an exponential gap between both decision diagrams, as the *BMD representation of the pseudo–Boolean function $2^x : \{0,1\}^n \to \mathbb{Z}$ defined by

$$\forall(\alpha_0, \ldots, \alpha_{n-1}) \in \{0,1\}^n : \; 2^x(\alpha_0, \ldots, \alpha_{n-1}) = 2^{\sum_{i=0}^{n-1} \alpha_i 2^i}$$

consists of a linear number of nodes whereas a BMD representation contains an exponential number of nodes. Figure 3.15a and Figure 3.15b show both representation for $n = 3$. It follows that with respect to efficient representations it is advantageous to use *BMDs.

4.2 Complexity of the essential operations

Now, let us investigate the complexities of the essential operations for *BMDs. We describe key algorithms for constructing and manipulating *BMDs. The algorithms have a similar style to their counterparts for ROBDDs. Unlike operations on ROBDDs where the complexities are at worst polynomial in the argument sizes, most operations on *BMDs potentially have exponential complexity. However, Bryant and Chen [22] showed that normally these exponential cases do not arise in applications.

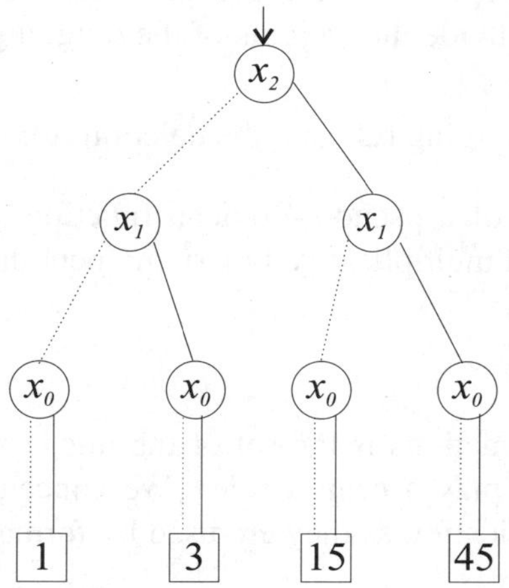

Figure 3.15a. BMD of function 2^x

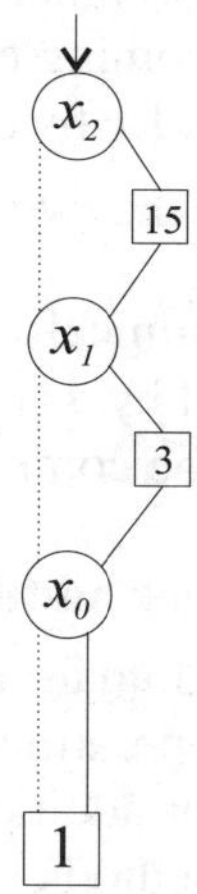

Figure 3.15b. *BMD of function 2^x

4.2.1 Maintaining canonicity

Given an ordered multiplicative binary moment diagram which does not meet the normalizing rules given above, we can transform it in a *BMD which fulfills the properties by a bottom-up pass through the diagram. For each node v, we first ask whether v is a terminal node or an internal node.

- If v is a terminal node labelled with a value $c \notin \{0, 1\}$, each incoming edge e of v is redirected to terminal node **1**. Additionally, the weight ω_e is multiplied by value c.

- If v is an internal node, we apply the following transformation steps:

  1. If the weight of the low-edge e of v is value 0, then edge e is redirected to terminal node **0**.

  2. If the low-edge e points to terminal node **0**, then the weight of e is set to 1. Additionally, the weight of the incoming edges of v are multiplied by the weight of the high-edge of v which is finally set to 1, too.

  3. If the weight of the low-edge e of v is a negative value, the weights of the incoming edges of v, the weight of the low-edge of v, and the weight of the high-edge of v are multiplied by -1.

  4. Compute the greatest common divisor

$$\omega_v := \gcd(\omega_{(v,low(v))}, \omega_{(v,high(v))})$$

of the weights of the low- and high-edge of v. Multiply the weights of the incoming edges of v by ω_v and divide the weights of the outgoing edges of v by ω_v.

5. Apply the two reduction rules, the merging rule and the deletion rule.

Thus, the canonical *BMD representation of a pseudo–Boolean function f which is given by a (non-canonical) ordered multiplicative binary moment diagram F can be computed in time $\mathcal{O}(|F|)$.

4.2.2 Integer addition

Since the co-domain of pseudo–Boolean functions is the set of the integers, arithmetical operations applied to *BMDs play a central role. We concentrate on integer addition and integer multiplication as they are used by formal verification methods.

We start with integer addition for BMDs. The basic algorithm for integer addition of BMDs which is shown in Figure 3.16 follows the ITE-algorithms for ROBDDs of Figure 3.8. It is based on the property that taking moments of pseudo–Boolean functions commutes with integer addition, i.e.,

$$\begin{aligned} f+g &= (f_{x_i=0} + x_i \cdot f_{\dot{x}_i}) + (g_{x_i=0} + x_i \cdot g_{\dot{x}_i}) \\ &= (f_{x_i=0} + g_{x_i=0}) + x_i \cdot (f_{\dot{x}_i} + g_{\dot{x}_i}) \end{aligned}$$

holds for all pseudo–Boolean functions f and g which are defined over the same set X_n of variables.

Since in the algorithm shown in Figure 3.16 the *smallest* variable x occurring in F and G is taken in each iteration, the computation of the BMDs $F_{x=0}$ and $F_{\dot{x}}$ of the constant moment $\phi(F)_{x=0}$ and linear moment $\phi(F)_{\dot{x}}$ of $\phi(F)$, respectively, can be performed in linear time (see Section 4.2.5).

Because of the computed table which is used by the ADD-algorithm, there are no more calls than pairs consisting of one node of F and one node of G. Thus, the overall algorithm for integer addition of two BMDs has worst case time complexity $\mathcal{O}(|F| \cdot |G|)$.

The algorithm for integer addition of *BMDs is slightly more complicated because of the edge weights. Let F and G be two *BMDs with root nodes r_F and r_G. Let us denote the weights of the incoming edges of r_F and r_G by ω_F and ω_G. Then, the following recursion holds

$$\begin{aligned} &\phi(F) + \phi(G) \\ &= \phi(\omega_F, r_F) + \phi(\omega_G, r_G) \\ &= \omega_F \cdot (\phi(\omega_{(r_F, low(r_F))}, low(r_F)) + x_i \cdot \phi(\omega_{(r_F, high(r_F))}, high(r_F))) + \\ &\quad \omega_G \cdot (\phi(\omega_{(r_G, low(r_G))}, low(r_G)) + x_i \cdot \phi(\omega_{(r_G, high(r_G))}, high(r_G))) \\ &= \omega_F \cdot \phi(\omega_{(r_F, low(r_F))}, low(r_F)) + \omega_G \cdot \phi(\omega_{(r_G, low(r_G))}, low(r_G)) + \end{aligned}$$

```
ADD(F,G)
begin

  if TERMINAL_CASE("ADD",F,G,result)
  then return result;
  fi;

  if FIND_IN_COMPUTED_TABLE("ADD",F,G,result)
  // true iff ADD(F,G) has already been computed
  then return result;
  fi;

  x = smallest variable occurring in F and G;
  low = ADD(F_{x=0}, G_{x=0});
  high = ADD(F_ẋ, G_ẋ);

  if high==0
  then result = low;
  else result = FIND_OR_ADD_UNIQUE_TABLE(x,low,high);
  fi;

  INSERT_IN_COMPUTED_TABLE("ADD",F,G,result);
  return result;

end;
```

Figure 3.16. Algorithm for adding two BMDs F and G ordered wrt. the same variable order.

$$
\begin{aligned}
& x_i \cdot (\omega_F \cdot \phi(\omega_{(r_F, high(r_F))}, high(r_F)) + \omega_G \cdot \phi(\omega_{(r_G, high(r_G))}, high(r_G))) \\
= {} & \phi(\omega_F \cdot \omega_{(r_F, low(r_F))}, low(r_F)) + \phi(\omega_G \cdot \omega_{(r_G, low(r_G))}, low(r_G)) + \\
& x_i \cdot (\phi(\omega_F \cdot \omega_{(r_F, high(r_F))}, high(r_F)) + \phi(\omega_G \cdot \omega_{(r_G, high(r_G))}, high(r_G))).
\end{aligned}
$$

Thus, the number of calls of the corresponding ADD procedure is not restricted anymore by the product of the number of nodes but by the product of the number of paths in the *BMDs under consideration on which the edge weights are "gathered". The number of paths of a *BMD can be exponential in the size of the operands.

4.2.3 Integer multiplication

Integer multiplication applied to two BMDs F and G representing the pseudo–Boolean functions f and g is based on the recursive equation

$$
\begin{aligned}
f \cdot g &= (f_{x_i=0} + x_i \cdot f_{\dot{x}_i}) \cdot (g_{x_i=0} + x_i \cdot g_{\dot{x}_i}) \\
&= f_{x_i=0} \cdot g_{x_i=0} + x_i \cdot (f_{x_i=0} \cdot g_{\dot{x}_i} + f_{\dot{x}_i} \cdot g_{x_i=0} + f_{\dot{x}_i} \cdot g_{\dot{x}_i}).
\end{aligned}
$$

The corresponding algorithm is shown in Figure 3.17.

```
MUL(F,G)
begin

  if TERMINAL_CASE("MUL",F,G,result)
  then return result;
  fi;

  if FIND_IN_COMPUTED_TABLE("MUL",F,G,result)
  // true iff MUL(F,G) has already been computed
  then return result;
  fi;

  x = smallest variable occurring in F and G;
  E1 = MUL(F_{x=0}, G_{x=0});
  E2 = MUL(F_{x=0}, G_ẋ);
  E3 = MUL(F_ẋ, G_{x=0});
  E4 = MUL(F_ẋ, G_ẋ);
  E5 = ADD(E2, E3);
  E6 = ADD(E5, E4);

  if E6==0
  then result = E1;
  else result = FIND_OR_ADD_UNIQUE_TABLE(x,E1,E6);
  fi;

  INSERT_IN_COMPUTED_TABLE("MUL",F,G,result);
  return result;

end;
```

Figure 3.17. Algorithm for multiplying two BMDs F and G.

The algorithm for *BMDs is a straightforward generalization of the algorithm shown in Figure 3.17. Its worst case time complexity is exponential in the size of the operands as the algorithm contains integer additions of *BMDs.

4.2.4 Boolean operations

The Boolean functions of $\mathcal{B}_n$ are just a special case of the numeric functions of $\mathcal{Z}_n$. Therefore, Boolean functions can be represented as *BMDs as well. For that purpose, we have to translate Boolean operations into arithmetical operations. This can be done in a quite easy way by using the relations

$$\begin{aligned} f \wedge g &= f \cdot g, \\ f \vee g &= f + g - f \cdot g, \end{aligned}$$

and

$$f' = 1 - f$$

for Boolean functions f and g of $\mathcal{B}_n$. (For the sake of clearness, we exceptionally use the symbols $\wedge$ and $\vee$ for denoting the logical-and operator and the logical-or operator, respectively, and the symbols $\cdot$ and + for denoting integer multiplication and integer addition, respectively.) It is easy to check that the minimum sized *BMD of the logical-or of multiple variables contains more internal nodes than the corresponding ROBDDs [23]. Thus, *BMDs do not guarantee the same representation size as ROBDDs. Actually, there are Boolean functions in $\mathcal{B}_n$ which have ROBDDs of polynomial size but *BMDs with exponential size only [4].

4.2.5 Moment computation

In *BMD based formal verification procedures, cofactor computation is not of interest but computation of the constant moment and linear moment of a pseudo–Boolean function with respect to a variable x_i. Usually, the smallest variable of the variable order of the *BMD or the smallest variable occurring in the *BMD is chosen for the decomposition.

Let F be a BMD. Let x_i be the first variable in the variable order of F. Then the BMDs $F_{x_i=0}$ and $F_{\dot{x}}$ of the constant moment $\phi(F)_{x_i=0}$ and linear moment $\phi(F)_{\dot{x}_i}$ of $\phi(F)$, respectively, are computed as follows: Either the root node r_F of BMD F is not labelled with x_i or it is labelled with x_i. In the first case, $\phi(F)$ is independent of x_i. Thus, $\phi(F)_{x_i=0} = \phi(F)$ holds and $\phi(F)_{\dot{x}_i}$ is the constant function $\mathbf{0}$. In the latter case, we have $\phi(F)_{x_i=0} = \phi(low(r_F))$ and $\phi(F)_{\dot{x}_i} = \phi(high(r_F))$.

The generalization to *BMDs F with root node r_F is straightforward because of

$$\phi(F)_{x_i=0} = \begin{cases} \phi(F) & \text{, if } index(r_F) \neq x_i \\ \phi(\omega_F \cdot \omega_{(r_F, low(r_F))}, low(r_F)) & \text{, if } index(r_F) = x_i \end{cases}$$

and

$$\phi(F)_{\dot{x}_i} = \begin{cases} \mathbf{0} & \text{, if } index(r_F) \neq x_i \\ \phi(\omega_F \cdot \omega_{(r_F, high(r_F))}, high(r_F)) & \text{, if } index(r_F) = x_i. \end{cases}$$

4.2.6 Substitution by functions

The substitution of variables in *BMDs by pseudo–Boolean functions plays a central role in the formal verification method of Hamaguchi et.al. [57], which will be explained in Chapter 4.

```
SUBST(F,x,G)
begin

  if TERMINAL_CASE("SUBST",F,x,G,result)
  then return result
  fi;

  if FIND_IN_COMPUTED_TABLE("SUBST",F,x,G,result)
  // true iff SUBST(F,x,G) has already been computed
  then return result
  fi;

  y = smallest variable occurring in F;

  if y == x
  then result = ADD(F_{x=0},MUL(G,F_ẋ));
  else
    E1 = SUBST(F_{y=0},x,G);
    E2 = SUBST(F_ẏ,x,G);
    E3 = FIND_OR_ADD_UNIQUE_TABLE(y,0,1);
    // returns the *BMD of the Boolean function y
    E4 = MUL(E2,E3);
    result = ADD(E1,E4);
  fi;

  INSERT_IN_COMPUTED_TABLE("SUBST",F,x,G,result);
  return result;

end;
```

Figure 3.18. Sketch of the substitution algorithm for *BMDs.

Let us consider the substitution of a variable x_i in the *BMD F of a pseudo–Boolean function f by a *BMD G of another pseudo–Boolean function g. According to the moment decomposition

$$f = f_{x_i=0} + x_i \cdot f_{\dot{x}_i}$$

of f with respect to x_i it is based on the formula

$$(f_{x_i=g}) = f_{x_i=0} + g \cdot f_{\dot{x}_i}.$$

A sketch of the substitution algorithm can be seen in Figure 3.18. It starts with the *BMD F, the *BMD G, and the variable x that is to be substituted. If the top variable y of F is not the variable to be substituted, the algorithm calls itself recursively with the low- and high-successor of the root of F. Then, the *BMD is reassembled . If variable x is found, an integer multiplication and an integer addition over *BMDs are performed according to the above equation.

The problem of the substitution algorithm is that the algorithms ADD for integer addition and MUL for integer multiplication have exponential worst case behavior when using *BMDs. However, under specific conditions which exist for example during multiplier verification by backward construction one can formally prove that the worst case will not occur [74]. We will deal with this fact in Chapter 4.

4.2.7 Minimization of *BMDs

Optimizing *BMDs has to be performed very carefully in order to avoid long running times. The problem is that exchanging neighboring variables in *BMDs does not only affect the neighboring levels under consideration [45]. Look at Figure 3.19 which shows two *BMDs of the pseudo–Boolean function f which is described by the arithmetic formula $x_1 - 2 \cdot x_2 + 2 \cdot x_1 \cdot x_2$ (cf. [45]). The

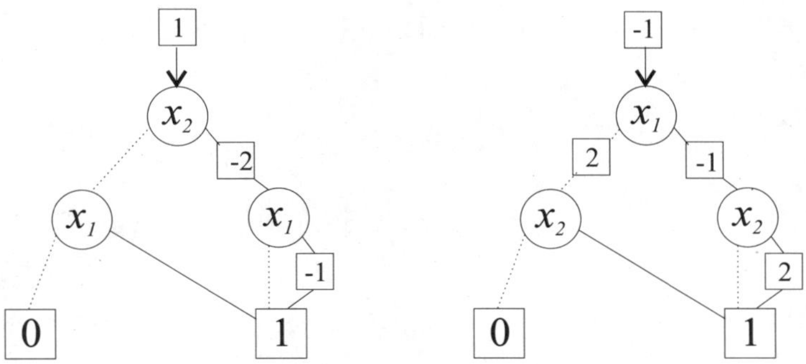

Figure 3.19. Two *BMDs of the same pseudo–Boolean function under different variable orders.

order of the right *BMD can be obtained by swapping both variables in the left *BMD only. Both *BMDs meet the normalizing rules. However, note that the two weights of the incoming edges of the root nodes differ. Now, imagine that this incoming edge is a low-edge of some internal node. Then, weights in upper levels have to be changed in order to restore the normalizing rules and maintain canonicity.

That is why, sifting should be applied to *BMDs without taking care on the normalizing rules. In a postprocessing step, the resulting *BMD is normalized in order to guarantee canonicity.

It is worthwhile to mention that Bryant and Chen [22] have noticed in their experiments that *BMDs are much less sensitive to variable ordering than ROBDDs.

Notes

1 The symbol $\cdot$ denotes both, the logical-and operator and the integer multiplication operator. In general, the meaning of the symbol will be clear from the context. If this is not the case (as for example in Section 4.2.4) we exceptionally use the symbol $\wedge$ for denoting the logical-and operator.

2 The symbol $+$ denotes both, the logical-or operator and the integer-addition operator. In general, the meaning of the symbol will be clear from the context. If this is not the case (as for example in Section 4.2.4) we exceptionally use the symbol $\vee$ for denoting the logical-or operator.

II

EQUIVALENCE CHECKING OF COMBINATIONAL CIRCUITS

In the second part of the book we concentrate on formal equivalence checking of combinational circuits. Chapter 4 reviews the straightforward approaches (Bryant 1986 [20], Bryant and Chen 1995 [23], Hamaguchi et.al. [57]) to combinational equivalence checking which consist of *transforming* the two combinational circuits under consideration *into canonical forms* and checking whether the two canonical forms are equal. Usually, ROBDDs and *BMDs are used in this context. Unfortunately, these approaches have some bottlenecks:

- most papers published in literature only handle the problem of proving that an error-free circuit is error-free,
- the representational power of ROBDDs and *BMDs is too weak to efficiently represent some important Boolean functions, and
- the memory requirements may grow exponentially during the synthesis of the decision diagram despite the application of efficient variable ordering heuristics. This can be the case even when the functional behavior of the Boolean network under consideration can efficiently be represented by an ROBDD or a *BMD.

We review the discussion about the first point made by Wefel and Molitor [145] and the approach presented by Hulgaard et.al [64] and Hett [60] which attack the second and the third difficulty.

Another approach to equivalence checking of combinational circuits is based on *automatic test pattern generation* (ATPG) and *satisfiability of conjunctive normal forms* (SAT). The common idea of both approaches is to transform the equivalence checking problem to a satisfiability problem. Chapter 5 attends to this SAT- and ATPG-based methods [16, 55, 86, 123]. In particular, we review the basic algorithms for ATPG and SAT [34, 35, 96, 97, 149].

Usually, the circuits which have to be compared contain a significant number of internal signal lines which are functionally equivalent. These *structural similarities* can be exploited in order to either simplify the *miter* (see Figure 5.1 on page 100) so that the corresponding satisfiability problem becomes easier or partition the problem into smaller and simpler equivalence checking problems. Chapter 6 presents both approaches [16, 77]. They are based on an idea of Berman and Trevillyan [9].

Chapter 7 addresses a special equivalence checking problem, the problem of *black box equivalence checking*. It occurs when the specification is known, but only parts of the implementation are finished or known. Black box equivalence checking enables the use of verification techniques in early stages of the design. Design errors can be already detected when only a partial implementation is at hand by combining parts of the implementation which are not yet finished into black boxes. The chapter has been written by the guest authors, Prof. Dr. Bernd Becker and Prof. Dr. Christoph Scholl, University Freiburg. It follows [129, 130].

Part II closes with the discussion of *permutation independent Boolean comparison*. This problem occurs in equivalence checking of combinational circuits since many verifiers use syntactical name matching to determine corresponding input variables and corresponding output variables of the circuits which have to be compared. This is one of the most important reasons for false negative results as during synthesis huge hierarchical designs lead to long hierarchical names for signal lines and backend tools often have severe restrictions on name lengths so that the names are shortened accordingly. Thus, the correspondence between the inputs and the outputs of the circuits under comparison has to be restored before calling equivalence checking methods as those presented in Chapters 4-6. Chapter 8 reviews the ideas presented in [105, 106, 108, 109].

Chapter 4

USE OF CANONICAL DATA STRUCTURES

The straightforward approach to solving the combinational logic-level equivalence problem is to use canonical forms. Here, the task of verifying whether two combinational circuits $\mathcal{N}_1$ and $\mathcal{N}_2$ are functionally equivalent consists of constructing canonical forms of the functions realized by $\mathcal{N}_1$ and $\mathcal{N}_2$. However, traditional canonical representations such as truth tables and minterm forms (see Chapter 3) are quite impractical. The truth table of every Boolean function with n arguments is of size 2^n. Many common functions applied in practice such as 32-bit integer addition and parity computation have no minterm form of practical size. ROBDDs and *BMDs are often substantially more compact than these traditional normal forms. For this reason, ROBDDs and *BMDs are usually used if canonical data structures have to be applied. However, since some Boolean functions cannot be efficiently represented by them either, we have to resort to other approaches for those Boolean functions. These approaches are going to be discussed in the subsequent chapters.

1. Synthesizing ROBDDs and *BMDs

In this section, we concentrate on the techniques available to create the ROBDD or the *BMD of the function realized by a Boolean network $\mathcal{N}$ with n inputs and m outputs. If we handle the primary outputs independently of each other, then we create an ROBDD F_i for each primary output y_i. We refer to this approach as *bit-level verification*. If the output variables are considered to represent an integer according to some encoding, then the Boolean network realizes a pseudo–Boolean function. Some common encodings are the unsigned binary, the sign-magnitude, and the two's complement – for a formal definition of these encodings we refer to Section 1.2.2. This approach is called *word-level verification*. Figure 4.1a and 4.1b illustrate this differentiation.

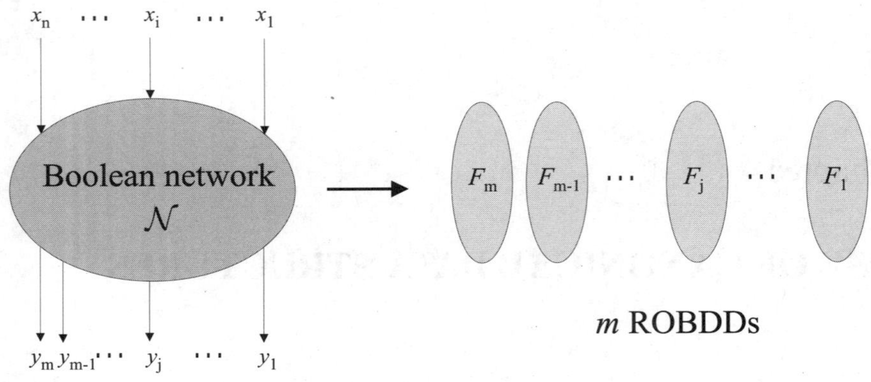

Figure 4.1a. Bit-level representation of a Boolean network $\mathcal{N}$. F_j denotes the ROBDD of the Boolean function computed at the jth primary output y_j.

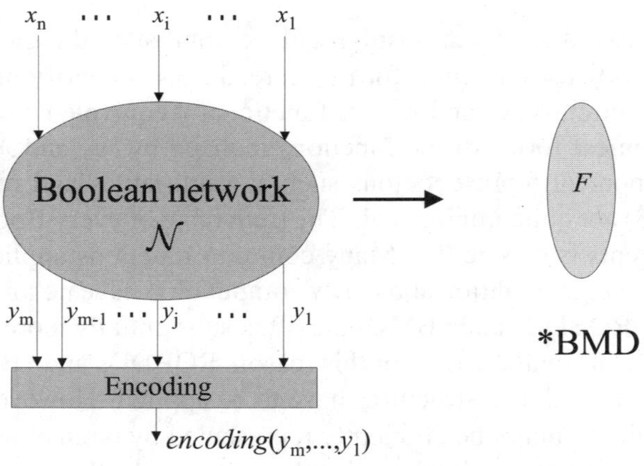

Figure 4.1b. Word-level representation of a Boolean network $\mathcal{N}$. F denotes the *BMD of the pseudo–Boolean function $f \in \mathcal{Z}_n$ defined by $encoding \circ (f_{y_m} \times \ldots \times f_{y_1})$.

1.1 Equivalence checking at the bit level

In the following, let us assume that the Boolean networks under consideration are defined over the standard cell library

$$\Omega_{std} = \{NOT, \cdot, \bar{\cdot}, +, \overline{+}, \oplus, \overline{\oplus}\} \subset \mathcal{B}_1 \cup \mathcal{B}_2.$$

Let $\mathcal{N}$ be such a Boolean network with n binary-valued inputs and one binary-valued output. (Note that it is enough to consider Boolean networks with one primary output only since, at the bit level, the outputs of a multi-output Boolean network have to be handled independently of each other.) Then, the

ITE-operator introduced in Chapter 3 Section 3 provides the basic method for creating the ROBDD representation of $\phi(\mathcal{N}) \in \mathcal{B}_n$.

At first, the algorithm computes a topological sort [31] of the nodes of the Boolean network $\mathcal{N}$. Then, it explores the nodes according to this topological sort. At each node v of $\mathcal{N}$ but the primary output it creates the ROBDD F_v of the Boolean function which is computed at the fanout edges of v. If v is the ith primary input, we associate v with the ROBDD F_v of the Boolean function $\boldsymbol{x}_i \in \mathcal{B}_n$. If v is an inner node, ROBDD F_v is computed by applying the ITE algorithm to the ROBDD(s) of the direct predecessor(s), accordingly to the following case differentiation with respect to the type of node v:

$$F_v = \begin{cases} ITE(F_{w_1}, \mathbf{0}, \mathbf{1}) & \text{, if } type(v) = \text{NOT} \\ ITE(F_{w_1}, F_{w_2}, \mathbf{0}) & \text{, if } type(v) = \cdot \\ ITE(F_{w_1}, ITE(F_{w_2}, \mathbf{0}, \mathbf{1}), \mathbf{1}) & \text{, if } type(v) = \bar{\cdot} \\ ITE(F_{w_1}, \mathbf{1}, F_{w_2}) & \text{, if } type(v) = + \\ ITE(F_{w_1}, \mathbf{0}, ITE(F_{w_2}, \mathbf{0}, \mathbf{1})) & \text{, if } type(v) = \overline{+} \\ ITE(F_{w_1}, ITE(F_{w_2}, \mathbf{0}, \mathbf{1}), F_{w_2}) & \text{, if } type(v) = \oplus \\ ITE(F_{w_1}, F_{w_2}, ITE(F_{w_2}, \mathbf{0}, \mathbf{1})) & \text{, if } type(v) = \overline{\oplus}. \end{cases}$$

If v is the primary output, i.e., $pa(1) = v$, then v has one incoming edge (w, v) for some node w only and F_v is set to F_w. Thus, $F_{pa(1)}$ is an ROBDD representation of the Boolean function $\phi(\mathcal{N})$ realized by the Boolean network $\mathcal{N}$.

The procedure just described is called *symbolic simulation* or *conventional symbolic simulation*, in literature.

1.2 Equivalence checking at the world level

In verifying arithmetic circuits, we usually abstract from the bit-level representation of a Boolean network, where each signal is binary-valued, to a word level, where bundles of signals encode words of data. The basic idea has been illustrated in Figure 4.1b.

1.2.1 Hierarchical approach

The hierarchical approach to equivalence checking of arithmetic circuits has been presented by Bryant and Chen [22, 23]. It is based on partitioning the circuits into components with easy word-level specifications. At first, it is shown that the bit-level implementation of a component module implements correctly its word-level specification. Then the composition of the word-level specifications which is derived accordingly to the interconnection structure of the circuit is verified.

A first idea is to construct a *BMD for each of the bit-level outputs of the components, much as would be done with ROBDDs. An implementation of the binary Boolean operations applied to *BMDs has been given in Chapter 3 Section 4.2.

For illustration, look at a circuit which should represent a full-adder with three inputs x_1, x_2, and x_3 and two outputs s and c denoting the sum- and the carry-output, respectively. Component verification first computes the *BMD of the binary-valued output s and the *BMD of the binary-valued output c. In case the circuit under consideration actually realizes a full-adder, the *BMDs of the Boolean functions computed at the signal lines s and c represent the Boolean functions

$$sum(x_1, x_2, x_3) = x_1 \oplus x_2 \oplus x_3$$

and

$$carry(x_1, x_2, x_3) = (x_1 \wedge x_2) \vee ((x_1 \oplus x_2) \wedge x_3),$$

respectively. (For the sake of clearness, once again we exceptionally use the symbols $\wedge$ and $\vee$ for denoting the logical-and operator and the logical-or operator, respectively. The symbols $\cdot$ and $+$ denote integer multiplication and addition, respectively, in the following considerations.) Applying the equations

$$\begin{aligned} a \wedge b &= a \cdot b, \\ a \vee b &= a + b - a \cdot b, \end{aligned}$$

and

$$\begin{aligned} a \oplus b &= (a \wedge b') \vee (a' \wedge b) \\ &= (a \cdot (1 - b)) \vee ((1 - a) \cdot b) \\ &= (a - a \cdot b) \vee (b - a \cdot b) \\ &= (a - a \cdot b) + (b - a \cdot b) - (a - a \cdot b) \cdot (b - a \cdot b) \\ &= a + b - 2 \cdot a \cdot b - (a \cdot b - a \cdot b - a \cdot b + a \cdot b) \\ &= a + b - 2 \cdot a \cdot b \end{aligned}$$

which hold for all $a, b \in \{0, 1\}$, the corresponding pseudo–Boolean functions can be given by

$$\begin{aligned} sum(x_1, x_2, x_3) &= x_1 \oplus x_2 \oplus x_3 \\ &= (x_1 + x_2 - 2 \cdot x_1 \cdot x_2) \oplus x_3 \\ &= (x_1 + x_2 - 2 \cdot x_1 \cdot x_2) + x_3 \\ &\quad - 2 \cdot (x_1 + x_2 - 2 \cdot x_1 \cdot x_2) \cdot x_3 \\ &= x_1 + x_2 + x_3 - 2 \cdot x_1 \cdot x_2 - 2 \cdot x_1 \cdot x_3 \\ &\quad - 2 \cdot x_2 \cdot x_3 + 4 \cdot x_1 \cdot x_2 \cdot x_3 \end{aligned}$$

and

$$carry(x_1, x_2, x_3) = x_1 \cdot x_2 + x_1 \cdot x_3 + x_2 \cdot x_3 - 2 \cdot x_1 \cdot x_2 \cdot x_3.$$

The two *BMDs are shown in Figure 4.2a and Figure 4.2b.

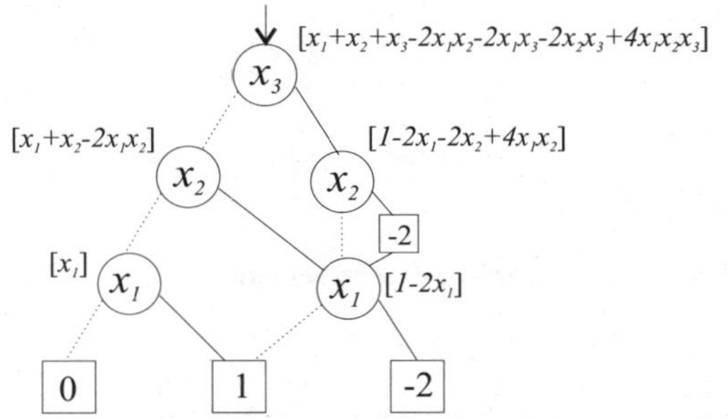

Figure 4.2a. *BMDs of the binary-valued output representing the sum-output s of a full-adder.

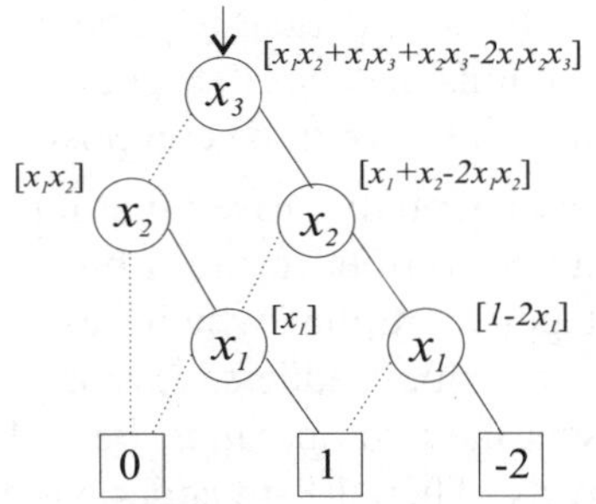

Figure 4.2b. *BMDs of the binary-valued output representing the carry-output c of a full-adder.

Now, the word-level representation of the components under consideration is obtained by simply generating the *BMD of $s + 2 \cdot c$ using integer multiplication and integer addition of *BMDs. Thus, we have transformed the bit-level representation of the components in a word-level representation of the circuit. Finally, the component verification is completed by comparing this word-level representation with the *BMD of the specification, which is the pseudo–Boolean function realized by a full-adder (see Figure 4.3).

For larger circuits, representing the bit-level functionality becomes too cumbersome and hence the method for component verification described above cannot be applied directly. For example, attempting to construct the bit-level functions of an integer multiplier (in particular the bit-level function realized by the middle bit of the multiplier) would cause an exponential blow-up with *BMDs [4, 29]. (Remember that the *BMD of the word-level representation of integer

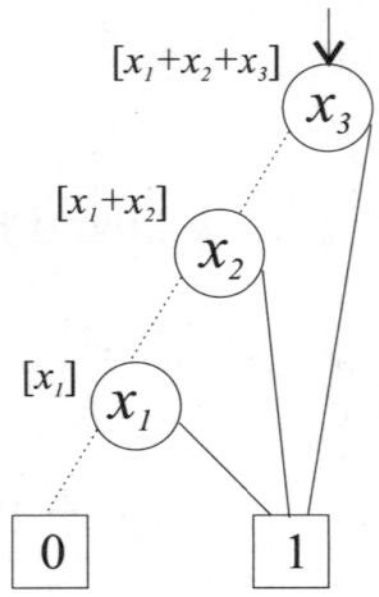

Figure 4.3. Word-level representation of a full-adder.

multiplication (see Figure 3.14b on page 56) has linear size.) Instead, Bryant and Chen follow a hierarchical approach in which the overall circuit is divided into components, each having a simple word-level specification. After the verification of the components the designer has to prove that the composition of the word-level component functions matches the specification. Mainly, the *BMD of an algebraic expression representing the composition is generated.

Bryant and Chen [23] experimentally proved that integer multipliers based on the add-step principle can be verified up to a bit width of 64 by using this approach. The n-bit integer multiplier which has been considered by them mainly consists of $n-1$ carry-ripple adders. Assume that the $2n$ primary input signal lines which represent the two operands are called $a_0, a_1, \ldots, a_{n-1}$ and $b_0, b_1, \ldots, b_{n-1}$, respectively. Then the ith adder has as input the multiplicand word A, which is given by $\sum_{i=0}^{n-1} a_i \cdot 2^i$ (a *BMD of such an unsigned encoding is shown in Figure 4.4a), the ith bit b_i of the multiplier, and a partial sum input word PI_i. It generates a partial sum word PO_i where the functionality is specified as

$$PO_i = PI_i + (b_i \cdot 2^i) \cdot A.$$

After having completed the component verification of the adders, the word-level specifications of the adders can be combined accordingly to the interconnection structure of the circuit. In this case this is an easy task since the encoding function for an adder input matches the output encoding of the preceding adder. The resulting algebraic expression is given by

$$0 + (b_0 \cdot 2^0 \cdot A) + \ldots + (b_i \cdot 2^i \cdot A) + \ldots + (b_{n-1} \cdot 2^{n-1} \cdot A)$$

for which a *BMD is generated. Because of the canonicity of *BMDs, this *BMD has to be the *BMD of the multiplication function (if the implementation of the multiplier under consideration indeed is error-free).

There are some problems arising when using this methodology. The first one is the need of high-level specifications of component modules. The designer has to exercise great care in hierarchically specifying the circuits in order to make the design suitable for hierarchical specification. Another problem is that once having changed from the bit-level to the word-level no information about the exact ordering of connections at module boundaries is accessible any longer. Bryant and Chen simply require that the encoding function for a component input has to match the output encoding of the component supplying that input [22]. However, there is no check whether this requirement is met.

1.2.2 Backward construction method

Taking the problems of the hierarchical approach into account, Hamaguchi et.al. [57] have proposed another *BMD based method for verifying arithmetic circuits which they called *verification by backward construction.* This method does not need any high-level information and it avoids the necessity of first computing a *BMD of each binary-valued output of the circuit under consideration.

As illustrated in Figure 4.1b, the primary outputs of a circuit are regarded as a word rather than a set of separate bits. In the case of arithmetic circuits, they are usually considered to represent an integer according to the unsigned binary, i.e.,

$$encoding(y_m, \ldots, y_1) = \sum_{i=1}^{m} y_i \cdot 2^{i-1}$$

the sign-magnitude, i.e.,

$$encoding(y_m, \ldots, y_1) = (-1)^{y_m} \cdot \sum_{i=1}^{m-1} y_i \cdot 2^{i-1},$$

or the two's complement encoding,

$$encoding(y_m, \ldots, y_1) = \left(\sum_{i=1}^{m-1} y_i \cdot 2^{i-1} \right) - y_m \cdot 2^{m-1}.$$

The *BMD which represents their interpretation as pseudo–Boolean function is constructed in the initialization step of the backward construction method. The *BMDs of the commonly used encodings are shown in Figure 4.4a-4.4c [23].

Now, the basic idea of the backward construction method is to put a cut across all primary outputs of the circuit. The cut is moved step by step towards the primary inputs according to any reverse topological sort [31] of the cells. After having moved a cell into the area between the cut and the primary outputs, the

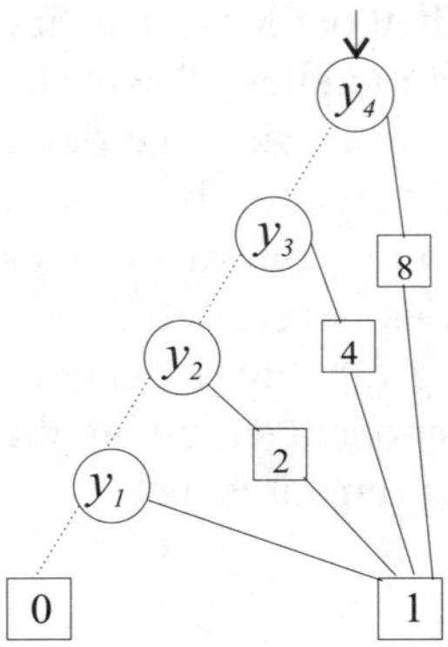

Figure 4.4a. *BMD of the unsigned binary encoding for m=4.

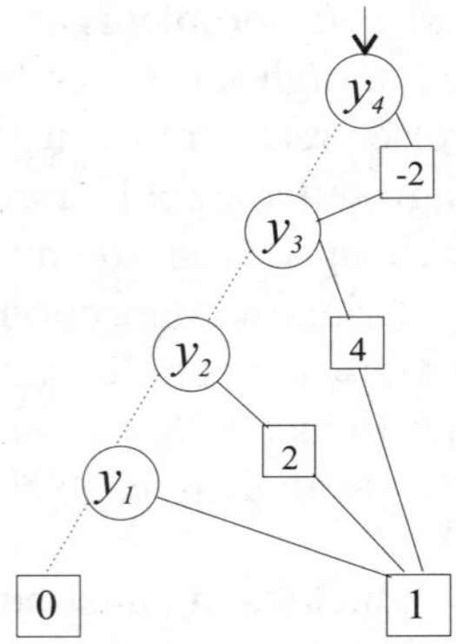

Figure 4.4b. *BMD of the sign-magnitude encoding for m=4.

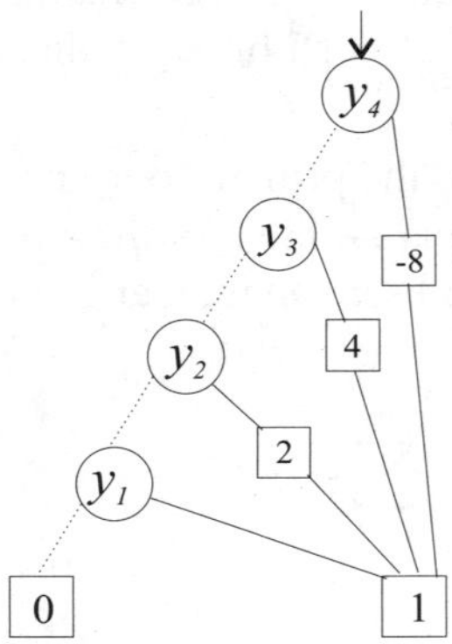

Figure 4.4c. *BMD of the two's complement encoding for m=4

*BMD of the subcircuit between the cut and the primary outputs is computed. Note that the *BMD of the word-level interpretation of the primary outputs corresponds to the cut which crosses all the primary outputs of the circuit. This *BMD is updated by substituting the outputs of the cell, that is moved, by the function realized by the cell. Here, the substitution algorithm presented in Chapter 3 Section 4.2 is used. Thus, after each iteration the *BMD only depends on the signals which are crossed by the cut. The algorithm stops if the cut crosses primary inputs only. It returns the *BMD belonging to this cut which obviously is the *BMD of the circuit under consideration.

Let us illustrate the approach of Hamaguchi et.al. by applying it to Wallace-tree like multiplier circuits [142]. Let $a = a_{n-1} \ldots a_0$ and $b = b_{n-1} \ldots b_0$ be two n-bit numbers in unsigned binary representation. The multiplication of a and b is equivalent to summing up the n *partial products* (of length $2n$) corresponding to the rows of the *partial product matrix P*:

$$P = \begin{pmatrix} 0 & 0 & \ldots & 0 & 0 & a_{n-1}b_0 & \ldots & a_1b_0 & a_0b_0 \\ 0 & 0 & \ldots & 0 & a_{n-1}b_1 & a_{n-2}b_1 & \ldots & a_0b_1 & 0 \\ \cdot & & \cdot & & & \cdot & & \cdot & \cdot \\ \cdot & & \cdot & & & \cdot & \cdot & & \cdot \\ \cdot & \cdot & & & & \cdot & \cdot & & \cdot \\ 0 & a_{n-1}b_{n-1} & \ldots & & & a_0b_{n-1} & 0 & \ldots & 0 \end{pmatrix}$$

A Wallace-tree like multiplier reduces the n partial products to two $2n$-bit integers s and c such that

$$s + c = (\sum_{i=0}^{n-1} a_i \cdot 2^i) \cdot (\sum_{j=0}^{n-1} b_j \cdot 2^j).$$

The reduction uses full-adders only, for example carry-save adders [116]. The depth of a Wallace-tree like multiplier depends on the arrangement of the full-adders. Ideally, it is $\mathcal{O}(\log n)$.

In the first step of the backward construction method, we have to encode the primary outputs of the Wallace-tree like multiplier. As the primary outputs represent two $2n$-bit integers which have to be added in order to finish the multiplication of the operands a and b, we generate the *BMD of the pseudo–Boolean function

$$\sum_{i=0}^{2n-1} s_i \cdot 2^i + \sum_{i=0}^{2n-1} c_i \cdot 2^i.$$

This *BMD has a specific structure which is called *Sum of weighted variables* (SOV) in [74]: In a *BMD in SOV, no two internal nodes are labelled with the same variable. Thus, there are at most as many internal nodes as variables. The high-edge of each internal node points to a terminal node. All low-edges but the low-edge of the last variable point to internal nodes and the low-edge of the last variable points to a terminal node. Thus, a *BMD over X_n in SOV represents a pseudo–Boolean function f of the form

$$f(x_1, \ldots, x_n) = w_0 + \sum_{i=1}^{n} w_i \cdot x_i$$

where $w_0, \ldots, w_n$ are integer values. w_i is called *weight* of variable x_i.

Let us return to the verification of our multiplier and let FA be a full-adder whose outputs are crossed by the cut. Let x_i and x_j be the sum- and carry-output of FA and x_k, x_l, and x_m the inputs of FA. It can be shown that the weight of the carry-output x_j is twice the weight of the sum-output x_i. The current BMD in SOV does not depend on x_k, x_l, or x_m. Now, we have to replace the variables x_i and x_j by the *BMDs of the functions $sum(x_k, x_l, x_m)$ and $carry(x_k, x_l, x_m)$, respectively (see Figure 4.5a) These replacements have to be executed one after

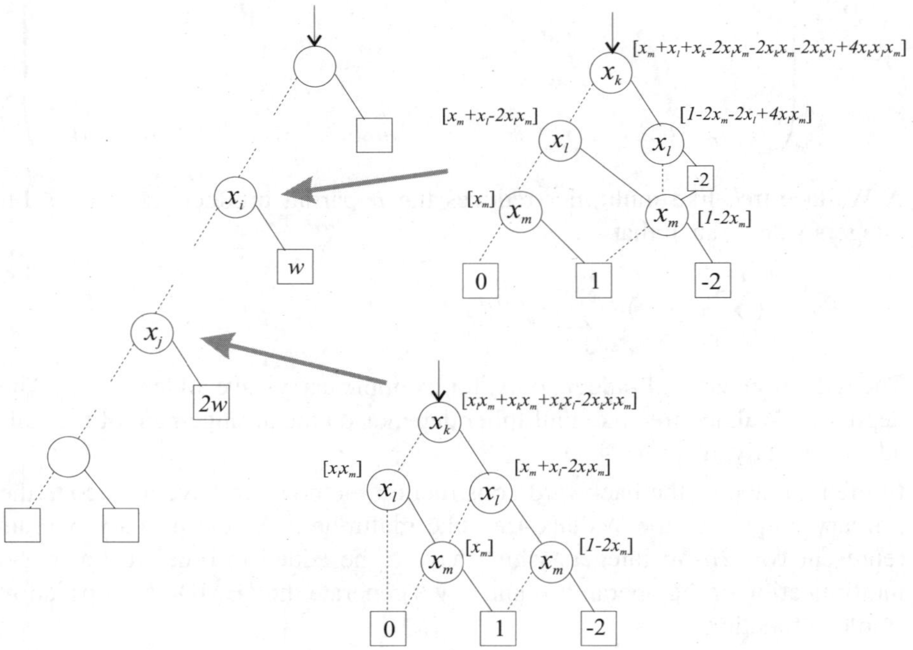

Figure 4.5a. Replacing the outputs of a full-adder by the corresponding *BMDs.

the other. Figure 4.5b shows the *BMD after substitution of x_i by the *BMD of $sum(x_k, x_l, x_m)$. In this *BMD, x_j is replaced by $carry(x_k, x_l, x_m)$. It is easy to see that the resulting *BMD is in SOV again, and that the variables x_k, x_l, and x_m have the same weight. Thus, the considerations just made can be repeated until all the full-adders have been moved to the other side of the cut. After the substitution of the partial product bits, we obtain the *BMD of the partial integer multiplication function. This provides that the circuit under verification is actually a Wallace-tree like multiplier.

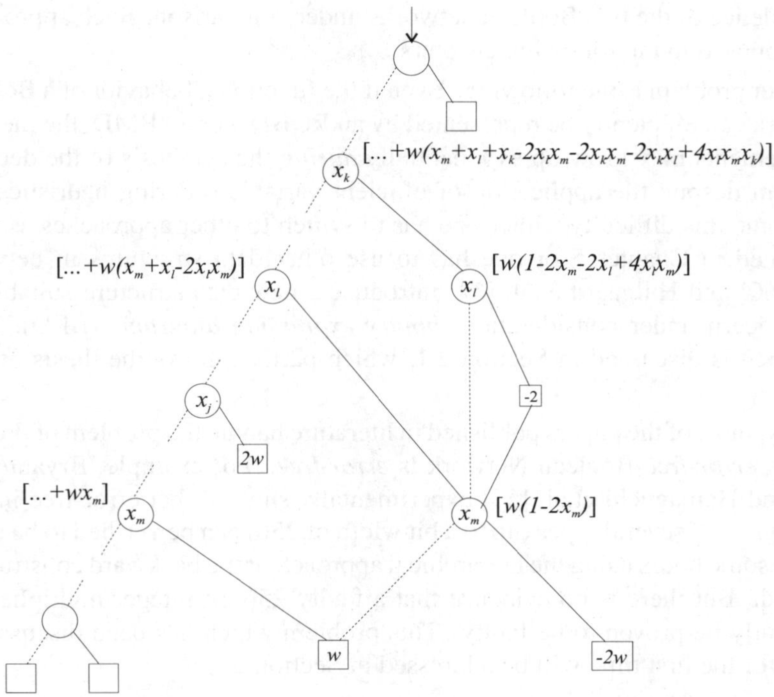

Figure 4.5b. *BMD after substitution of x_i by the *BMD of $sum(x_k, x_l, x_m)$.

A detailed analysis of the complexity of backward construction applied to Wallace-tree like multipliers can be found in [74]. Hamaguchi et.al. [57] experimentally proved that 256-bit Wallace-tree integer multipliers can be verified using this approach.

2. Problems arising when using one of these methodologies

The approaches to equivalence checking just presented have three major bottlenecks.

At first, some functions cannot be efficiently represented by ROBDDs or *BMDs. Bryant [20] proved that the ROBDDs for integer multipliers are of exponential size. The representational power of *BMDs is too weak to efficiently represent integer division [131]. For Boolean networks realizing such functions, equivalence checking which is solely based on canonical forms fails due to the exponential blow-up of the size of the data structures. In practice, once the size of the data structures exceeds a given boundary during a run, the algorithm is stopped and one has to switch to other approaches in order to decide the equivalence of the two Boolean networks under comparison. Such approaches are proposed in the following chapters.

Another problem is the following. Even if the functional behavior of a Boolean network can efficiently be represented by an ROBDD or a *BMD, the memory requirements may grow up exponentially *during* the synthesis of the decision diagram despite the application of efficient variable ordering heuristics. To overcome this difficulty, either one has to switch to other approaches as those presented in Chapter 5 or one has to use other data structures in between. Hett [60] and Hulgaard et.al. [64] introduce a new data structure suitable for the concern under consideration, *binary expression diagrams* (BED). Their approach is discussed in Section 2.1, which partly follows the thesis of Hett [60].

Finally, most of the papers published in literature handle the problem of proving that an *error-free* Boolean Network is *error-free*. For example, Bryant/Chen [23] and Hamaguchi et.al. [57] experimentally showed that error-free integer multipliers of several types (up to a bit width of 256) can be verified to be error-free in some hours using the hierarchical approach or the backward construction method. But there is no evidence that a faulty 256-bit integer multiplier can efficiently be proven to be faulty. This problem which has been discussed in [145] for the first time will be addressed in Section 2.2.

2.1 Avoiding intermediate size peaks

In order to verify that two Boolean networks $\mathcal{N}_1$ and $\mathcal{N}_2$ with n inputs and one output are equivalent, the classical approach described above constructs either the ROBDDs of $\phi(\mathcal{N}_1)$ and $\phi(\mathcal{N}_2)$ or the ROBDD of $\phi(\mathcal{N}_1) \Leftrightarrow \phi(\mathcal{N}_2)$. Due to the canonicity of ROBDDs, the two Boolean networks implement the same Boolean function if and only if the ROBDDs of $\phi(\mathcal{N}_1)$ and $\phi(\mathcal{N}_2)$ are equal or the ROBDD of $(\phi(\mathcal{N}_1) \Leftrightarrow \phi(\mathcal{N}_2))$ is identical to the ROBDD which consists of terminal node **1** only. In general, the procedure first computes the ROBDDs of $\phi(\mathcal{N}_1)$ and $\phi(\mathcal{N}_2)$ which are combined as $(\phi(\mathcal{N}_1) \oplus \phi(N_2))'$.

For illustration, look at the following example which we have taken from [60]. Let g_1, g_2, and g_3 be three Boolean functions of $\mathcal{B}_n$ which do not depend on the variables x_1 and x_2. Furthermore, let us assume, that for all $i, j \in \{1, 2, 3\}$ with $i \neq j$ the ROBDD of $g_i \cdot g_j$ is quadratic in the size of the ROBDDs of g_i and g_j. In order to synthesize the ROBDD F of

$$(x_1 \cdot g_1) \cdot (x_2 \cdot g_2) \cdot ((x_1 \oplus x_2) \cdot g_3)$$

the classical approach first computes the ROBDD of either

$$(x_1 \cdot g_1) \cdot (x_2 \cdot g_2),$$

$$(x_1 \cdot g_1) \cdot ((x_1 \oplus x_2) \cdot g_3),$$

or

$$(x_2 \cdot g_2) \cdot ((x_1 \oplus x_2) \cdot g_3).$$

Thus, the algorithm takes at least quadratic time before returning the terminal node **0** which is the final result F. If the ROBDDs of g_1, g_2, and g_3 are quite large, the consequence is a blow-up of the ROBDDs during the synthesis of the ROBDD F such that the algorithm aborts without success.

In order to possibly overcome such a blow-up of the size of the data structures, it would be helpful to have a generalization of BDDs which allows rearranging Boolean formulae. In the above example, a convenient rearrangement results in a quick transformation of the Boolean formulae to 0:

$$\begin{aligned}&(x_1 \cdot g_1) \cdot (x_2 \cdot g_2) \cdot ((x_1 \oplus x_2) \cdot g_3)\\ &= (x_1 \cdot x_2 \cdot (x_1 \oplus x_2)) \cdot (g_1 \cdot g_2 \cdot g_3)\\ &= 0 \cdot (g_1 \cdot g_2 \cdot g_3)\\ &= 0.\end{aligned}$$

In some sense, binary expression diagrams allow such rearrangements. They have a similar structure as BDDs (see Definition 3.6 on page 31) and are defined as follows.

DEFINITION 4.1 (BINARY EXPRESSION DIAGRAMS) *Let X_n be the set of variables $\{x_1, \ldots, x_n\}$ and $\bigodot := \{\odot_1, \ldots, \odot_p\}$ a set of binary Boolean operators. A binary expression diagram over $(X_n, \bigodot)$ is a directed acyclic graph $G = (V, E, label, value)$ with node set V containing two types of nodes, internal and terminal nodes. A terminal node v is labelled with $value(v)$ which is either 0 or 1. An internal node v has as label either a variable $x_i \in X_n$ or a binary Boolean operator $\odot_j \in \bigodot$, i.e., $label(v) \in X_n \cup \bigodot$, and two children $low(v)$ and $high(v)$. An internal node v is called operator node if $label(v) \in \bigodot$. It is called variable node if $label(v) \in X_n$.*

The relation between a binary expression diagram and the Boolean function it represents is straightforward.

DEFINITION 4.2 (INTERPRETATION OF BINARY EXPRESSION DIAGRAMS) *Let $G = (V, E, label, value)$ be a binary expression diagram over $(X_n, \bigodot)$ and v a node of V. The Boolean function $\phi(v) \in \mathcal{B}_n$ which is computed at node v is recursively defined as follows:*

- *If v is a terminal node and $value(v) = 1$, then $\phi(v) = \mathbf{1}$.*
- *If v is a terminal node and $value(v) = 0$, then $\phi(v) = \mathbf{0}$.*
- *If v is an internal node and $label(v) = x_i$, then*

$$\phi(v) = (x_i{}' \cdot \phi(low(v))) + (x_i \cdot \phi(high(v)))$$

where $\cdot$ and $+$ denote the logical-and operator and the logical-or operator, respectively.

- *If v is an internal node and $label(v) = \odot_j$, then*

$$\phi(v) = \phi(low(v)) \odot_j \phi(high(v)).$$

In the following we only consider reduced ordered binary expression diagrams with complemented edges (cf. Chapter 3 Section 3.3) over $(X_n, \bigodot)$ which we call BED in short. A BED is *ordered* if on all paths through the graph each variable occurs at most once and the variables respect a given total order. It is *reduced*, if neither the deletion rule nor the merging rule introduced for BDDs can be applied to a variable node, the operands of an operator node v are not the same nodes, i.e., $low(v) \neq high(v)$, and none of both operands of a operator node is a terminal node. Here we assume that $\bigodot = \{\cdot, \overline{\cdot}, +, \overline{+}, \oplus, \overline{\oplus}\}$. If $low(v) = high(v)$ holds for an operator node v with $label(v) = \odot_j$, it is easy to see that

$$\phi(v) = \begin{cases} \mathbf{0} & \text{, if } \odot_j = \oplus \\ \mathbf{1} & \text{, if } \odot_j = \overline{\oplus} \\ \phi(low(v)) & \text{, if } \odot_j \in \{+, \cdot\} \\ \phi(low(v))' & \text{, if } \odot_j \in \{\overline{+}, \overline{\cdot}\}. \end{cases}$$

If the low-successor (or the high-successor) of the operator node is a terminal node then the BED can be reduced as follows

$$\phi(v) = \begin{cases} \mathbf{0} & \text{, if } low(v) = \mathbf{0} \text{ and } \odot_j = \cdot \\ \mathbf{1} & \text{, if } low(v) = \mathbf{0} \text{ and } \odot_j = \overline{\cdot} \\ \phi(high(v)) & \text{, if } low(v) = \mathbf{0} \text{ and } \odot_j \in \{+, \oplus\} \\ \phi(high(v))' & \text{, if } low(v) = \mathbf{0} \text{ and } \odot_j \in \{\overline{+}, \overline{\oplus}\} \\ \mathbf{0} & \text{, if } low(v) = \mathbf{1} \text{ and } \odot_j \in \{\overline{+}\} \\ \mathbf{1} & \text{, if } low(v) = \mathbf{1} \text{ and } \odot_j \in \{+\} \\ \phi(high(v)) & \text{, if } low(v) = \mathbf{1} \text{ and } \odot_j \in \{\cdot, \overline{\oplus}\} \\ \phi(high(v))' & \text{, if } low(v) = \mathbf{1} \text{ and } \odot_j \in \{\overline{\cdot}, \oplus\}. \end{cases}$$

BEDs are an extension of ROBDDs with complemented edges. They are capable of representing any Boolean network with n inputs and one output in linear space and partly contain the structural composition of the Boolean network. The rearrangement of operators is performed by shifting operator nodes in the BED which is a local operation. For example, moving variable nodes above operator nodes can be done in a very efficient way as cofactor computation commutes with each of the binary Boolean operators. Look at Figure 4.6. It is

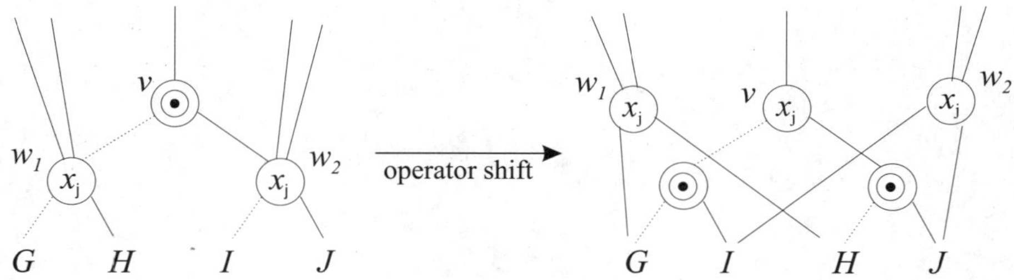

Figure 4.6. Shifting an operator node [60]. G, H, I, and J are sub–BEDs.

easy to see that

$$\begin{aligned}
\phi(v) &= ((x_j{}' \cdot G) + (x_j \cdot H)) \odot ((x_j{}' \cdot I) + (x_j \cdot J)) \\
&= (x_j{}' \cdot (((x_j{}' \cdot G) + (x_j \cdot H)) \odot ((x_j{}' \cdot I) + (x_j \cdot J)))_{x_j=0}) \\
&\quad +(x_j \cdot (((x_j{}' \cdot G) + (x_j \cdot H)) \odot ((x_j{}' \cdot I) + (x_j \cdot J)))_{x_j=1}) \\
&= (x_j{}' \cdot (((1 \cdot G) + (0 \cdot H)) \odot ((1 \cdot I) + (0 \cdot J)))) \\
&\quad +(x_j \cdot (((0 \cdot G) + (1 \cdot H)) \odot ((0 \cdot I) + (1 \cdot J)))) \\
&= (x_j{}' \cdot ((G + 0) \odot (I + 0))) + (x_j \cdot ((0 + H) \odot (0 + J))) \\
&= (x_j{}' \cdot (G \odot I)) + (x_j \cdot (H \odot J)).
\end{aligned}$$

Now, look at our example from above where we tried to synthesize the ROBDD F of

$$(x_1 \cdot g_1) \cdot (x_2 \cdot g_2) \cdot ((x_1 \oplus x_2) \cdot g_3).$$

Let us assume that the ROBDDs G_1, G_2, and G_3 of g_1, g_2, and g_3, respectively, are given. A BED which corresponds to this Boolean formula is shown in Figure 4.7a. Let us move the operator nodes towards the terminal nodes. Moving operator node v_1 across the successor node labelled with x_1 (remember that we have assumed that g_1, g_2, and g_3 do not depend on neither x_1 nor x_2) results in the BED of Figure 4.7b. Applying the BED specific reduction rules results in the BED shown in Figure 4.7c. Analogously, operator node v_2 can be moved downwards which is illustrated in Figure 4.7d. In the next step, the operator nodes v_3 and v_4 can be moved below variable x_1. The BED which we obtain after performing these two operator shifts is shown in Figure 4.7e.

The BEDs shown in Figure 4.7f, Figure 4.7g, and Figure 4.7h show the further process. Moving the upper operator node towards the terminal nodes results

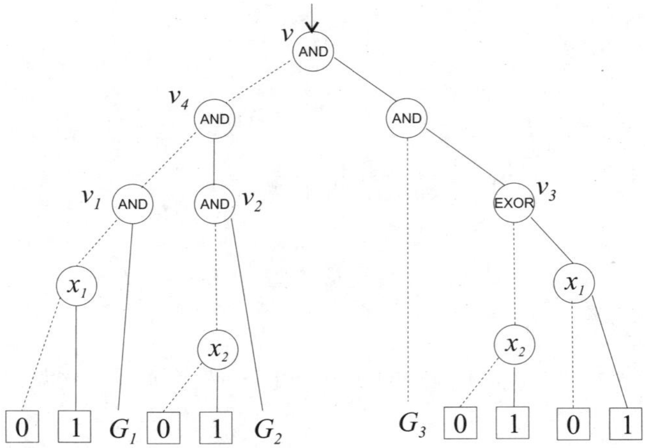

Figure 4.7a. BED of $(x_1 \cdot g_1) \cdot (x_2 \cdot g_2) \cdot ((x_1 \oplus x_2) \cdot g_3)$. For the sake of clarity, the merging rule has not been applied throughout the diagram.

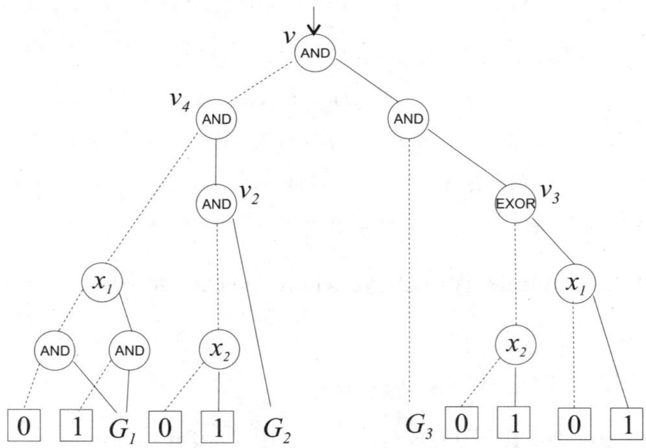

Figure 4.7b. BED after moving operator node v_1 across the successor node labelled with x_i.

into the *BMD shown in Figure 4.7h. This shows that BEDs allow to compute the BED of the Boolean formulae

$$(x_1 \cdot g_1) \cdot (x_2 \cdot g_2) \cdot ((x_1 \oplus x_2) \cdot g_3)$$

without exploring the ROBDD of g_1, g_2, or g_3.

The BED based verification procedure is quite simple. Given two Boolean networks $\mathcal{N}_1$ and $\mathcal{N}_2$ over $\mathcal{B}_1 \cup \mathcal{B}_2$ with n inputs and one output, we construct

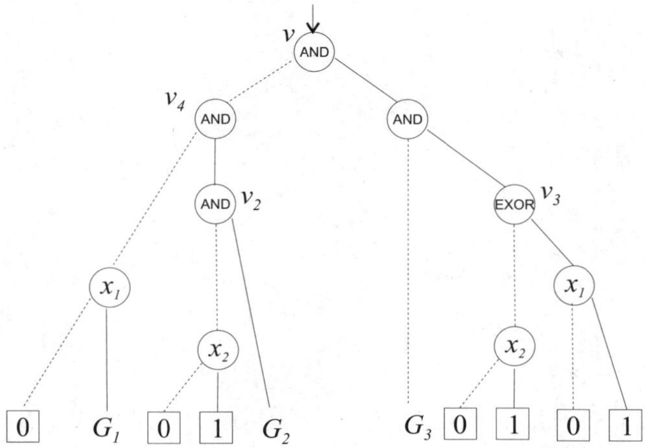

Figure 4.7c. Applying the BED specific reduction rules.

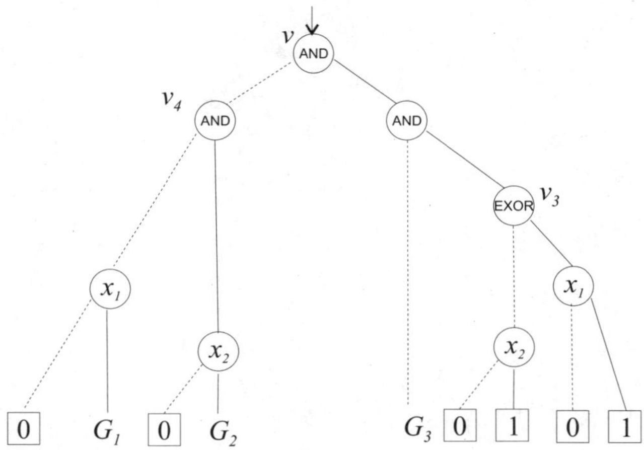

Figure 4.7d. Moving operator node v_2 downwards.

the BEDs B_1 and B_2 of $\mathcal{N}_1$ and $\mathcal{N}_2$, respectively, and combine them as $B_1 \oplus B_2$ by means of an operator node. Then, we convert this BED to a ROBDD by repeatedly moving operator nodes towards the terminal nodes. Operator nodes which point to terminal nodes are eliminated by the requirements of reduction. The worst time complexity of this transformation procedure can be exponential in the size of the BED as there are efficient BEDs representing

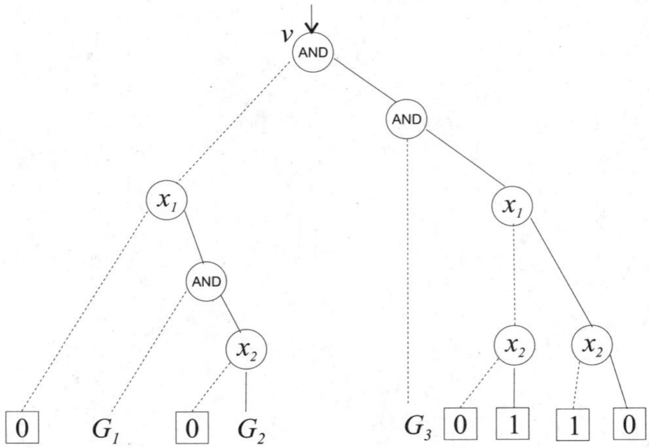

Figure 4.7e. Moving the operator nodes v_3 and v_4 towards the terminal nodes.

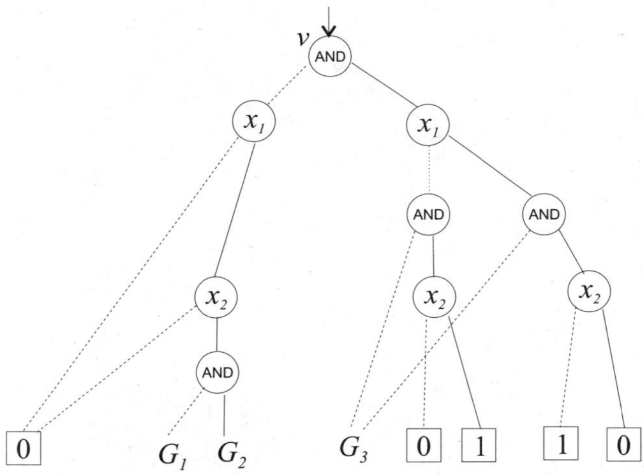

Figure 4.7f.

integer multiplication. However, in many cases BEDs are a suitable tool which helps avoiding intermediate size peaks during the synthesis of ROBDDs.

2.2 How to prove a faulty circuit to be faulty?

Now, let us come to the third problem of combinational equivalence checking based on canonical forms. Note that the investigations presented in the above sections deal with error-free implementations only. For example, the ap-

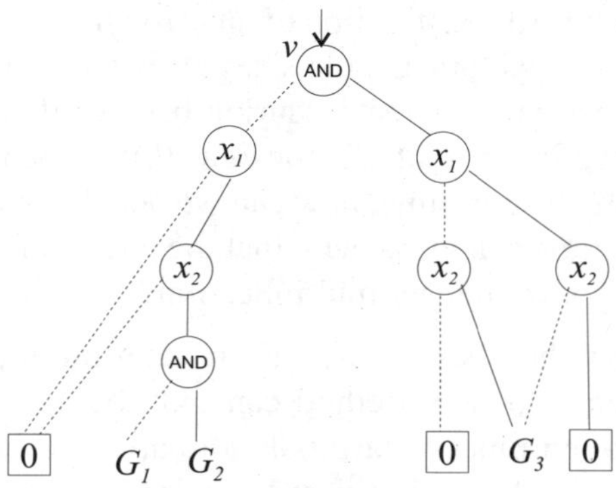

Figure 4.7g.

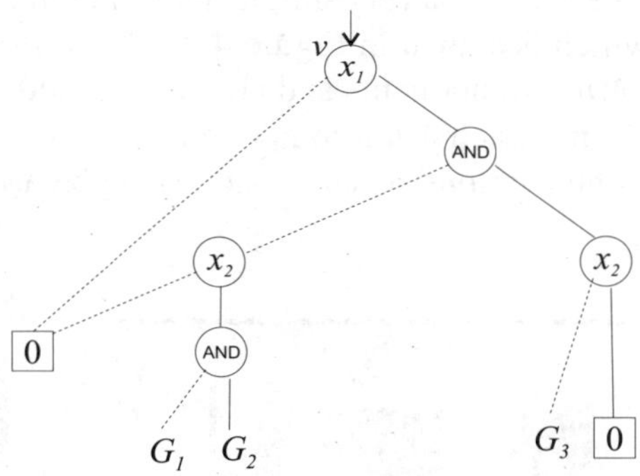

Figure 4.7h.

proaches of Bryant/Chen [23] and Hamaguchi et.al. [57] show by experiments that an error-free integer multiplier can be efficiently proven to be error-free. There is no evidence that a faulty integer multiplier can be proven to be faulty in polynomial time when using one of the approaches presented above.

In this chapter, we concentrate on the problem of how to formally prove that a faulty multiplier implementation is faulty. At a first glance, the answer of this question appears to be quite simple. Just build the *BMD of the pseudo–Boolean function realized by the circuit under consideration by applying the backward construction method and compare it to the *BMD of n-bit integer multiplication. Of course, if the circuit does not implement integer multiplication, the comparison fails. However, this drafted procedure only works if the *BMD of the pseudo–Boolean function realized by the faulty circuit can be synthesized in reasonable time. Up to now, we only know that the *BMDs occurring during backward construction of an *error-free* (Wallace-tree like) integer multiplier circuit have practicable sizes. It is quite possible that the size of the *BMDs exponentially increases during backward construction if some design error is inserted in the multiplier design. But, if the backward construction does not stop after a short time, how can we decide that the process can be aborted? After what time can we be sure that the circuit under consideration is faulty, i.e., does not realize integer multiplication?

We are going to show that these questions are relevant in practice. We prove that the backward construction method can only be applied as filter during formal logic verification unless sharp polynomial upper bounds for the sizes of the *BMDs occurring during backward construction of an error-free circuit of a particular type have been proven. Keim et.al. did it for Wallace-tree like integer multipliers [74].

To simplify matters, we consider a very simple multiplier implementation. It has the basic structure which is shown in Figure 4.8. The various partial products are computed in the *Mult*-components and are directly added to the subtotal in the next *Add*-component. Each *Mult*-component consists of AND-gates only. For the *Add*-components, various types of adders can be used.

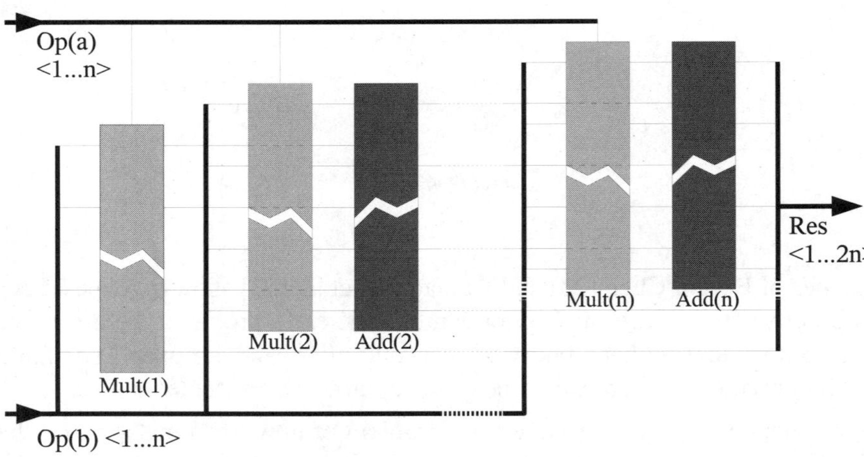

Figure 4.8. n-bit integer multiplier

We investigate the sizes of the *BMDs occurring during backward construction of an error-free implementation and various faulty implementations of our multiplier type. To obtain experimental data, let us construct by backward construction the *BMDs of these hardware implementations. In order to get comparable results, we are going to use the same breadth first search strategy to determine the order in which the substitutions are applied every time.

2.2.1 Error-free multiplier

Because of canonicity, the *BMD of the pseudo–Boolean function realized by an error-free n-bit integer multiplication hardware is isomorphic to the *BMD shown in Figure 3.14b on page 56. It contains $2 \cdot n$ inner nodes only. However, some *BMDs occurring during the backward construction of such an error-free multiplier are more complex. In particular, additional temporary variables have to be inserted in order to represent the subtotals of the multiplication and the various partial products. Figure 4.9 shows the sizes of the *BMDs occurring during the backward construction of an error-free 10-bit integer multiplier[1]. The data on the y-axis show the sizes of the intermediate *BMDs occurring during the corresponding substitution steps. As already described, the back-

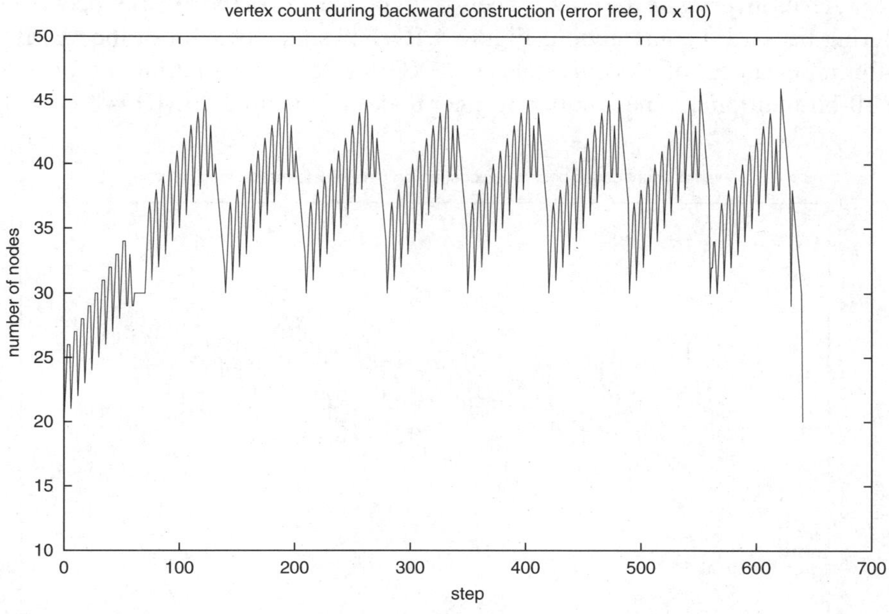

Figure 4.9. Backward construction of an error-free 10-bit multiplier

ward construction method applied to the n-bit integer multiplier starts with the *BMD of the output word which consists of the output variables $y_{2n-1}, \ldots, y_0$. Thus, in the case of a 10-bit multiplier the initial *BMD has size 20. During the first substitution steps the upper 10 output variables (see Figure 4.8) are substituted with respect to the function which is realized by the back adder block. Temporary variables for representing the subtotal of the multiplication

and the partial product which are operands of this back adder block have to be inserted. This results in a linear growth of the *BMDs (see Figure 4.9, step 0 to step 50). Substituting the output variables of the back multiplier block (step 50 to step 80) results in *BMDs with about 30 inner nodes. The resulting *BMDs after these steps only depend on output variables which have to be substituted later on, variables for representation of the subtotal, and some input variables. By having a close look at Figure 4.9 it is easy to recognize the various levels of our multiplier. During the substitutions due to an adder block the sizes of the *BMDs are approximately 45. After the substitutions due to a multiplier block the sizes of the *BMDs drop back to 30.
The experiments conducted for different bit widths show that the size of the *BMDs occurring during backward construction of our error-free n-bit integer multiplier are bounded by $6 \cdot n$.

2.2.2 Faulty multipliers: cell faults

Now, let us investigate the effect of cell faults on the size of the *BMDs occurring during backward construction. Figure 4.10 shows the behavior of the *BMDs after replacement of all AND-gates by EXOR-gates in the multiplier blocks of a 10-bit multiplier. The number of inner nodes of the final *BMD is $3 \cdot n - 1$.

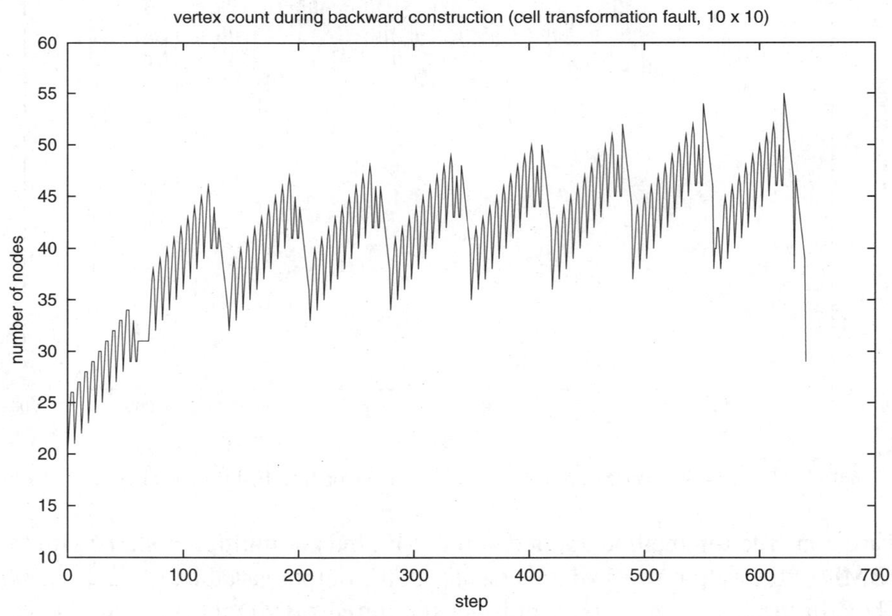

Figure 4.10. Backward construction of a faulty 10-bit multiplier (cell fault)

The sizes of the *BMDs occurring during backward construction also seem to be linear in the operand size. Thus, the *BMD of such a faulty multiplier can be constructed by backward construction even for large operands.

All the experiments pointed out that cell faults in our multiplier seem to be good-natured.

2.2.3 Faulty multipliers: wiring faults

The situation is rather precarious when the multiplier implementation contains a routing design error. Let us investigate two different wiring faults, a *hierarchical fault* which occurs in every level of the multiplier and a *local fault* where one connection is wrongly wired only.

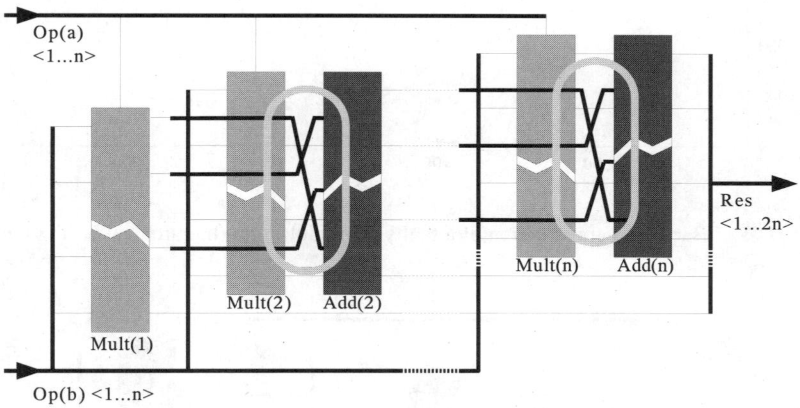

Figure 4.11. Local wiring fault

Hierarchical wiring fault. The first wiring fault which we investigate is a faulty routing between the outputs of the adder blocks which represent subtotals of the multiplication and the inputs of the next adder blocks. In particular, we shift (more exactly: we rotate) the subtotal by one position (see Figure 4.11 for illustration). Figure 4.12 shows the sizes of the *BMDs occurring during backward construction of such a faulty 6-bit multiplier implementation. The final size of the *BMD of the pseudo–Boolean function realized by this faulty 6-bit multiplier is still relatively small. It consists of 130 inner nodes. However, the *BMDs occurring during backward construction are very large, namely up to about 49.000 inner nodes. Due to memory restrictions, multipliers for larger operands cannot be considered.

Local wiring fault. The second wiring fault which we investigate is a faulty handling of the connection between the back adder block and the next to the

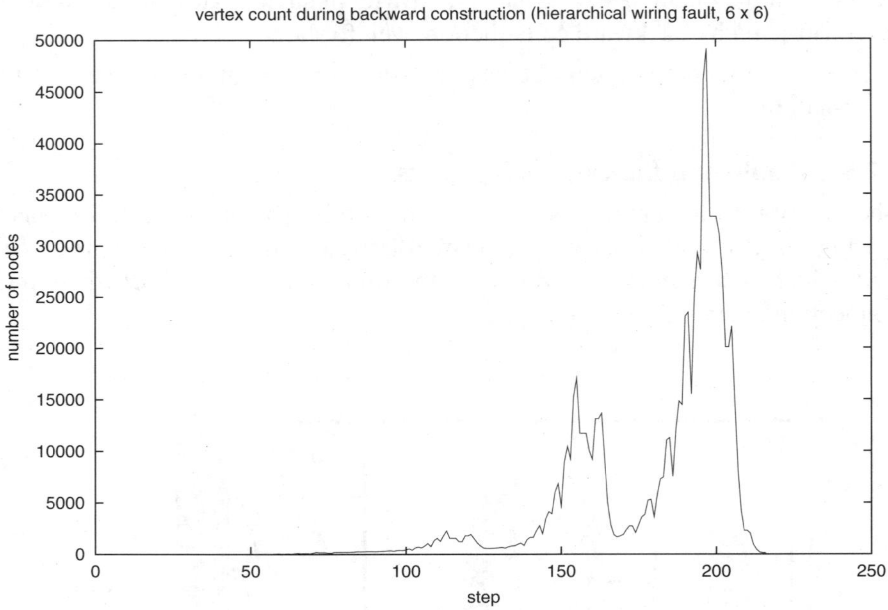

Figure 4.12. Backward construction of a faulty 6-bit multiplier (hierarchical wiring fault)

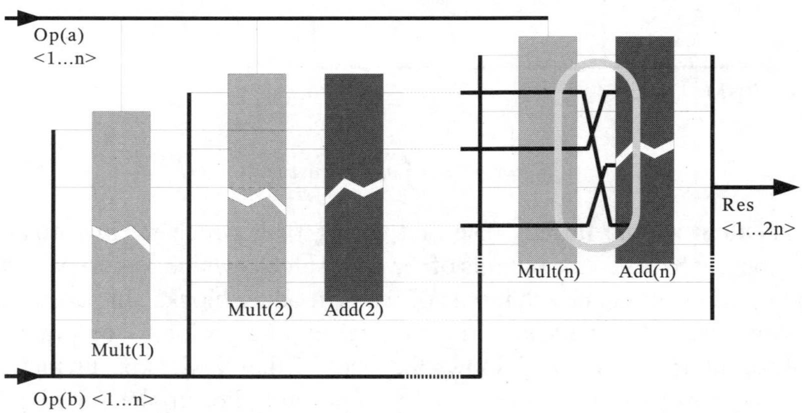

Figure 4.13. Local wiring fault

back adder block. Once again, we 'rotate' the subtotal. The fault is sketched in Figure 4.13. Figure 4.14 shows the behavior of the *BMDs for this case. The final *BMD size for $n = 7$ is 412. During the backward construction, *BMDs whose sizes are up to 125.000 nodes occur.

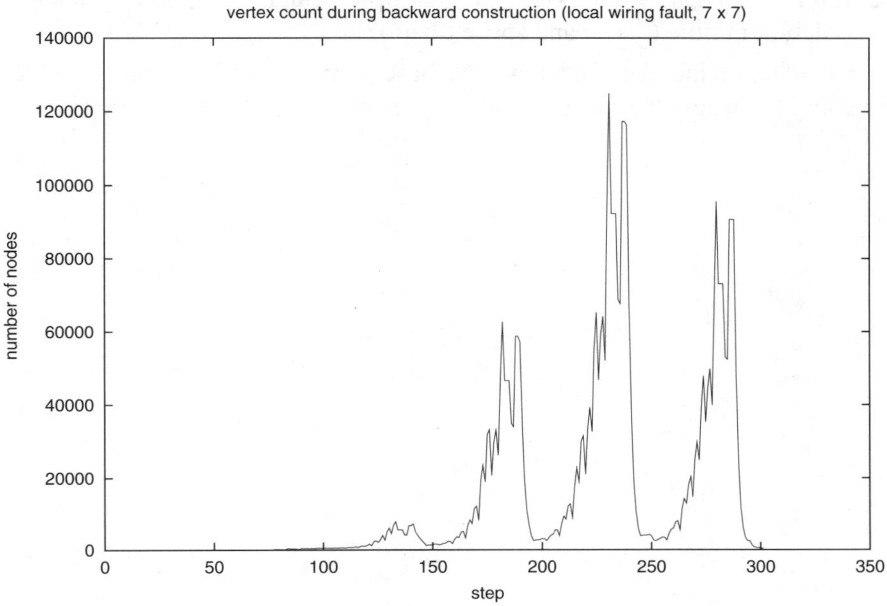

Figure 4.14. Backward construction of a faulty 7-bit multiplier (local wiring fault)

2.2.4 Conclusions

The considerations just made show that very simple faults can dramatically change the behavior of the backward construction method of Hamaguchi et al. Simple faults can result in very large *BMDs during backward construction although backward construction applied to an error-free implementation uses *BMDs of moderate sizes only. Consequently, the backward construction method can only be used as formal verification tool if polynomial upper bounds for the sizes of the *BMDs occurring during the backward construction applied to an *error-free* circuit of a particular type have been proven. Once the size of a *BMD exceeds this upper bound, we can conclude that the circuit under consideration does not belong to this particular circuit type. If such upper bounds have not been proven, the backward construction method can only be used as filter during formal logic verification. Only, if the method stops after a reasonable time, we have proven equivalence or non-equivalence. Otherwise, we do not know anything. The circuit under consideration may be error-free or faulty.

Note that asymptotic upper bounds are not of much help. Rather we need upper bounds with exact "constants". This approach was carried out for Wallace-tree like multipliers by Keim et. al. [74]. Here the constants needed can easily be pulled out of the constructive proof of the main theorem of the paper which states that any n-bit Wallace-Tree like multiplier can be verified by backward construction in time $\mathcal{O}(n^4)$ and space $\mathcal{O}(n^2)$.

If such upper bounds are not known, we have to resort to other approaches. We are going to discuss the most important ones in the subsequent chapters.

Notes

1 We used the *BMD-package of Chen which has already been used in [23]. In order to make it applicable for our purposes, we extended the *BMD package by the substitution function which has been presented in Chapter 3 Section 4.2 and by a counter which allows to record the size of the *BMDs during a substitution step. All tests were executed on a PC with 256 Megabyte main memory without additional swap memory.

Chapter 5

SAT AND ATPG BASED EQUIVALENCE CHECKING

Another approach to equivalence checking of combinational circuits is based on automatic test pattern generation (ATPG) and satisfiability of conjunctive normal forms (SAT). The common idea of both approaches is to transform the equivalence checking problem to a satisfiability problem. To realize this idea, an "exclusive-or" is applied to the two Boolean networks F and G which are to be compared. Figure 5.1 illustrates how to proceed. The term *miter* is used to refer to the configuration shown (cf. [16]). Thus, two Boolean networks F and G are functionally equivalent if and only if there is no assignment to the inputs of the miter of F and G such that the miter evaluates to 1. Thus, the following three statements are equivalent:

- F and G are not functionally equivalent;
- The miter of F and G is satisfiable, i.e., there is an assignment to the inputs of the miter of F and G such that the output evaluates to 1;
- The output of the miter of F and G is testable for stuck-at-0 (see [1]).

Of course, the BDD (or *BMD) based approach proposed in Chapter 4 can be used for checking whether the miter is satisfiable as well. Just construct the ROBDD of the miter of F and G. The miter is satisfiable if and only if its decision diagram does not only consist of terminal node **0**. This test can be done in constant time. However, the memory requirements for constructing the ROBDD of the miter may be too large and other approaches have to be applied. In this chapter we discuss approaches to combinational equivalence checking which apply traditional SAT algorithms and ATPG. In particular, we review the basic algorithms for automatic test pattern generation of combinational circuits and the satisfiability problem applied to conjunctive normal forms.

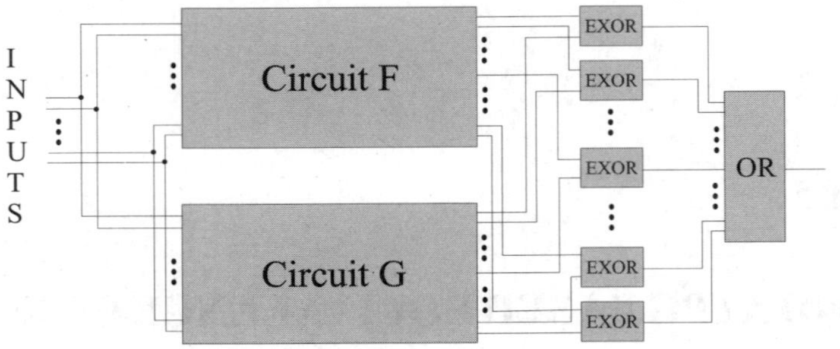

Figure 5.1. A miter.

1. SAT based equivalence checking

For applying SAT algorithms to the equivalence checking problem, the miter under consideration has to be transformed into a conjunctive normal form which characterizes the set of consistent assignments to the signal lines of the miter. Then, SAT algorithms are applied to this conjunctive normal form.

Conjunctive normal forms are particular Boolean formulae over X_n (cf. Definition 3.4).

DEFINITION 5.1 (CLAUSE) *A disjunction of literals, where a literal is the occurrence of a variable x_i or its complement $x_i{}'$, is called clause.*

In the following, we assume that a clause c does not contain both, literal L and literal L' and that the literals in a clause are unique. So we can treat clauses as sets of literals. With $|c|$, we are going to denote the number of literals contained in clause c. The empty set $\emptyset$ is called *empty clause* which is a description of the constant Boolean function $\mathbf{0}$.

DEFINITION 5.2 (CONJUNCTIVE NORMAL FORM) *A conjunction of clauses is called conjunctive normal form.*

Analogously to clauses, conjunctive normal forms can be treated as sets of clauses. Thus, $|\chi|$ denotes the number of clauses of the conjunctive normal form χ. The empty set $\emptyset$ is called *empty conjunctive normal form* and describes the constant Boolean function $\mathbf{1}$.

Despite the satisfiability problem applied to conjunctive normal forms being an NP-complete decision problem [53], SAT algorithms have seen dramatic improvements in recent years.

1.1 Generating a conjunctive normal form

As mentioned above, the miter $\mathcal{N}$ is transformed into a conjunctive normal form first [86]. For this purpose every edge of the Boolean network $\mathcal{N}$ has an associated variable. For illustration, please see Figure 5.2 which we have taken from [83].

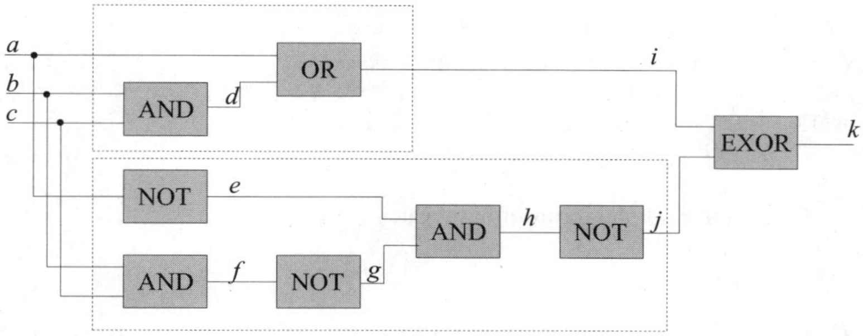

Figure 5.2. Example of a miter [83].

Each node of $\mathcal{N}$ is tagged with a conjunctive normal form that represents the function performed by the appropriate gate. We assume that the node v under consideration has two inputs x_1 and x_2 and one output z, i.e., $type(v) \in \mathcal{B}_2$ (cf. Definition 3.3). Then an assignment of the node v is called *consistent* if and only if it satisfies the formula

$$z = type(v)(x_1, x_2).$$

This formula is logically equivalent to

$$(z \Rightarrow type(v)(x_1, x_2)) \cdot (type(v)(x_1, x_2) \Rightarrow z)$$

and

$$(z' + type(v)(x_1, x_2)) \cdot (type(v)(x_1, x_2)' + z).$$

It is rather easy to transform such a formula into a conjunctive normal form for a particular node v of the Boolean network. For example, if v represents an AND-gate, i.e., $type(v) = \cdot$, then the Boolean formula tagged to v is given by

$$(z' + (x_1 \cdot x_2)) \cdot ((x_1 \cdot x_2)' + z)$$

which is logically equivalent to

$$(z' + x_1) \cdot (z' + x_2) \cdot (z + {x_1}' + {x_2}').$$

This formula states that whenever output z is 1, then the inputs x_1 and x_2 also have to be 1 as well; otherwise the first two clauses are not satisfied.

Because of the third clause, input x_1 or input x_2 has to be 0 if z is 0. Table 5.1 [86] summarizes the conjunctive normal forms of some other basic gates. It is straightforward to generalize these Boolean formulae to multi-input basic

$type(v)$	**conjunctive normal form**
$\cdot$	$(z' + x_1) \cdot (z' + x_2) \cdot (z + x_1{}' + x_2{}')$
$+$	$(z + x_1{}') \cdot (z + x_2{}') \cdot (z' + x_1 + x_2)$
$\bar{\cdot}$	$(z + x_1) \cdot (z + x_2) \cdot (z' + x_1{}' + x_2{}')$
$\overline{+}$	$(z' + x_1{}') \cdot (z' + x_2{}') \cdot (z + x_1 + x_2)$
$\oplus$	$(z' + x_1 + x_2) \cdot (z' + x_1{}' + x_2{}') \cdot (z + x_1{}' + x_2) \cdot (z + x_1 + x_2{}')$
NOT	$(z' + x_1{}') \cdot (z + x_1)$

Table 5.1. Conjunctive normal forms of basic gates

gates. For example, the consistent assignments of an n-input AND-gate $z = AND(x_1, \ldots, x_n)$ are characterized by

$$\prod_{i=1}^{n}(z' + x_i) \cdot (z + \sum_{i=1}^{n} x_i{}')$$

where $\prod$ and $\sum$ denote the iterative logical-and operator and the iterative logical-or operator, respectively.

Since each node is tagged with a Boolean formula which has to be independently satisfied, we can extract a characteristic formula $\chi(\mathcal{N})$ for any circuit $\mathcal{N}$ by taking the conjunction of all of the formulas of nodes in $\mathcal{N}$. The resulting conjunctive normal form describes the set of the consistent assignments to the signals of $\mathcal{N}$, i.e., an assignment which is consistent for every node in $\mathcal{N}$. For example, the characteristic formula $\chi(\mathcal{N})$ of the Boolean Network $\mathcal{N}$ shown in Figure 5.2 is

$$\begin{aligned} \chi(\mathcal{N}) \;=\;& (d' + b) \cdot (d' + c) \cdot (d + b' + c') \\ & \cdot\; (e' + a') \cdot (e + a) \\ & \cdot\; (f' + b) \cdot (f' + c) \cdot (f + b' + c') \\ & \cdot\; (g' + f') \cdot (g + f) \\ & \cdot\; (h' + e) \cdot (h' + g) \cdot (h + e' + g') \\ & \cdot\; (i + a') \cdot (i + d') \cdot (i' + a + d) \\ & \cdot\; (j' + h') \cdot (j + h) \\ & \cdot\; (k' + i + j) \cdot (k' + i' + j') \cdot (k + i' + j) \cdot (k + i + j'). \end{aligned}$$

Now, remember that in our application the Boolean network $\mathcal{N}$ we consider is a miter of some circuits F and G. Assume that k is the variable associated to the primary output of $\mathcal{N}$. Then, expanding $\chi(\mathcal{N})$ by clause k results in a formula $\chi(\mathcal{N}) \cdot k$ which is satisfiable if and only if the combinational circuits F and G are not functionally equivalent.

1.2 Algorithms for SAT-solving

Most algorithms for solving the SAT problem for conjunctive normal forms are based on the Davis-Putnam procedure [35] and Davis-Logemann-Loveland procedure [34]. In the following, let us follow the original paper of Davis and Putnam [35].

1.2.1 Davis-Putnam / Davis-Logemann-Loveland procedure

The *Davis-Putnam procedure* (DP) and *Davis-Logemann-Loveland procedure* (DLL) is based upon the iterative application of several rules, namely

- two *termination rules*,
- two *Boolean Constraint Propagation rules* (BCP),
  - the *one-literal clause rule*
  - the *monotone variable fixing rule*, and
- the *variable elimination rule.*

Termination rules. The termination rules which are applied are the following:

Termination rule I if χ is the empty conjunctive normal form, then χ is satisfiable.

Termination rule II if χ contains an empty clause, then χ is not satisfiable.

In the algorithm presented in Figure 5.3 on page 107, procedure TERMINAL_CASE returns TRUE, if termination rule I could be applied, FALSE, if termination rule II could be applied, or UNSOLVED. In the latter case, the SAT procedure has to apply BCP or the variable elimination rule.

One-literal clause rule. The *one-literal clause rule* (or *unit clause rule*) states that if a conjunctive normal form χ contains a clause c which consists of only one literal L, then χ can be modified by striking out all clauses that contain the literal L and deleting all occurrences of the negated literal L' from the remaining clauses of $\mathcal{N}$.

LEMMA 5.1 (ONE-LITERAL CLAUSE RULE) *Let χ be a conjunctive normal form and let χ_{new} be the conjunctive normal form which results if the one-literal clause rule is applied to χ. Then, χ_{new} is satisfiable if and only if χ is satisfiable.*

PROOF: The correctness of this rule is due to the equivalences $L \cdot (L + B) = L$ and $L \cdot (L' + B) = L \cdot B$, B being any Boolean formula. ∎

Let us illustrate the one-literal clause rule by applying it to the above example. Obviously, we can eliminate variable k in $\chi(\mathcal{N}) \cdot k$:

$$\begin{aligned} \phi(\chi(\mathcal{N}) \cdot k) \neq \mathbf{0} \quad \Longleftrightarrow \quad & \phi((d' + b) \cdot (d' + c) \cdot (d + b' + c') \\ & \cdot (e' + a') \cdot (e + a) \\ & \cdot (f' + b) \cdot (f' + c) \cdot (f + b' + c') \\ & \cdot (g' + f') \cdot (g + f) \\ & \cdot (h' + e) \cdot (h' + g) \cdot (h + e' + g') \\ & \cdot (i + a') \cdot (i + d') \cdot (i' + a + d) \\ & \cdot (j' + h') \cdot (j + h) \\ & \cdot (i + j) \cdot (i' + j')) \neq \mathbf{0} \,. \end{aligned}$$

Monotone variable fixing rule. The *monotone variable fixing rule* (*positive-negative rule*, or *pure literal rule*) resembles the above elimination rule. It claims that if variable x_i occurs in the conjunctive normal form χ only either as positive or negative literal, then all clauses which contain variable x_i may be deleted.

LEMMA 5.2 (MONOTONE VARIABLE FIXING RULE) *The resulting conjunctive normal form is satisfiable if and only if χ is satisfiable.*

PROOF: Without loss of generality, let us assume that none of the clauses of χ contains the negative literal $x_i{}'$. Then χ can be written as $\chi = D \cdot I$ where D and I are conjunctive normal forms. D consists of the clauses of χ which contain variable x_i. I consists of the clauses which are independent of x_i. Since I does not contain variable x_i, D can be satisfied by setting x_i to 1 without affecting the satisfiability of I. Thus, χ is satisfiable if and only if I is satisfiable. ∎

Variable elimination rule. The *variable elimination rule* or *splitting rule* eliminates variables in conjunctive normal forms. Let the given conjunctive normal form χ be put into the Boolean formula

$$(A + x_i) \cdot (B + x_i{}') \cdot I,$$

where A, B, and I are conjunctive normal forms which are independent of variable x_i. A and B are obtained by grouping together the clauses of χ that contain variable x_i and factoring out the occurrences of x_i and the occurrences of $x_i{}'$, respectively. I consists of the remaining clauses of χ. Then the following two logically equivalent rules hold.

LEMMA 5.3 (VARIABLE ELIMINATION RULE) *χ is not satisfiable if and only if $(A + B) \cdot I$ is not satisfiable.*

LEMMA 5.4 (SPLITTING RULE) *χ is not satisfiable if and only if $A \cdot I$ is not satisfiable and $B \cdot I$ is not satisfiable.*

PROOF: χ is not satisfiable, i.e., $\phi(\chi) = \mathbf{0}$, if and only if both the positive cofactor $\phi(\chi)_{x_i=1}$ and the negative cofactor $\phi(\chi)_{x_i=0}$ of $\phi(\chi)$ with respect to x_i are not satisfiable. Because of

$$\phi(\chi)_{x_i=1} = \phi(B \cdot I) \text{ and } \phi(\chi)_{x_i=0} = \phi(A \cdot I),$$

the equivalence

$$\phi(\chi) = \mathbf{0} \iff \phi(A \cdot I) = \mathbf{0} \text{ and } \phi(B \cdot I) = \mathbf{0}$$

holds. Thus, χ is satisfiable if and only if $A \cdot I$ is satisfiable or $B \cdot I$ is satisfiable, i.e., if and only if

$$\phi((A \cdot I) + (B \cdot I)) \neq \mathbf{0}$$

holds. ∎

Although theoretically both the variable elimination rule and the splitting rule are equivalent, usually the splitting rule is used since the variable elimination rule can easily increase the number and the lengths of the clauses in the expression after several applications [34, 37, 49].

A simple implementation of DLL is shown in the next section. However, before going into more details, we illustrate DLL by applying it to the above conjunctive normal form $\chi(\mathcal{N}) \cdot k$ which has already been reduced by the elimination of one-literal clauses. It is easy to see that BCP is not applicable. Thus, we have to apply the splitting rule. So, let d be the variable we want to eliminate in the next iteration. Then

$$\begin{aligned}
A &= (b' + c') \cdot (i' + a) \\
B &= (b) \cdot (c) \cdot (i) \\
I &= (e' + a') \cdot (e + a) \\
&\quad \cdot\ (f' + b) \cdot (f' + c) \cdot (f + b' + c') \\
&\quad \cdot\ (g' + f') \cdot (g + f) \\
&\quad \cdot\ (h' + e) \cdot (h' + g) \cdot (h + e' + g') \\
&\quad \cdot\ (i + a') \\
&\quad \cdot\ (j' + h') \cdot (j + h) \\
&\quad \cdot\ (i + j) \cdot (i' + j').
\end{aligned}$$

Thus, to prove satisfiability of $\chi(\mathcal{N}) \cdot k$, we have to check whether $\phi(A \cdot I) \neq \mathbf{0}$ or $\phi(B \cdot I) \neq \mathbf{0}$. Let us start with $B \cdot I$ where BCP can be applied. Eliminating the three one-literal clauses $\{b\}$, $\{c\}$, and $\{i\}$ of the conjunctive normal form $B \cdot I$ results in

$$\begin{aligned}
\phi(B \cdot I) \neq \mathbf{0} \quad \Longleftrightarrow \quad & \phi((e' + a') \cdot (e + a) \cdot (f) \\
& \cdot (g' + f') \cdot (g + f) \\
& \cdot (h' + e) \cdot (h' + g) \cdot (h + e' + g') \\
& \cdot (j' + h') \cdot (j + h) \cdot (j')) \ \neq \mathbf{0}.
\end{aligned}$$

The next iteration of BCP eliminates the one-literal clauses $\{f\}$ and $\{j'\}$. Thus

$$\begin{aligned}
\phi(B \cdot I) \neq \mathbf{0} \quad \Longleftrightarrow \quad & \phi((e' + a') \cdot (e + a) \cdot (g') \cdot (h' + e) \\
& \cdot (h' + g) \cdot (h + e' + g') \cdot (h)) \ \neq \mathbf{0}
\end{aligned}$$

holds. By eliminating the one-literal clause $\{g'\}$ in the next iteration, the right side of the above equivalence reduces to the conjunctive normal form

$$(e' + a') \cdot (e + a) \cdot (h' + e) \cdot (h') \cdot (h),$$

which is not satisfiable because there are both, the one-literal clause $\{h\}$ and the one-literal clause $\{h'\}$. This fact is detected by DLL by eliminating clause $\{h'\}$ which results in the conjunctive normal form

$$(e' + a') \cdot (e + a) \cdot (),$$

which contains the empty clause $\emptyset$.

Thus, $\chi(\mathcal{N}) \cdot k$ is satisfiable if and only if $A \cdot I$ is satisfiable.

Before closing the section we will stress that for most SAT solvers a major portion (about 90% in most cases) of the solver's run time is spent in the BCP process [112, 136]. Therefore, an efficient BCP engine is the key to any efficient SAT solver.

1.2.2 Implementation of the Davis-Logemann-Loveland procedure

An overall Davis-Logemann-Loveland algorithm is shown in Figure 5.3. The algorithm is called by SATISFY(χ). In the BCP phase, the one-literal clause

```
SATISFY(χ)
begin
  switch BCP(χ)
  case SUCCESS:     return SUCCESS;
  case FAILURE:     return FAILURE;
  case UNSOLVED:    return SPLIT(χ);
end;


BCP(χ)
begin

  if TERMINAL_CASE(χ,result) then return result fi;

  while χ can be simplified by a Boolean constraint rule
  do simplify χ;
     if TERMINAL_CASE(χ,result) then return result fi;
  od;

  return UNSOLVED;

end;


SPLIT(χ)
begin
  choose a literal L occurring in χ;
  if SATISFY(χ · L) then return SUCCESS else return SATISFY(χ · L') fi;
end;
```

Figure 5.3. Davis-Logemann-Loveland procedure

rule and the monotone variable fixing rule are iteratively applied to the conjunctive normal form χ. After each step, procedure TERMINAL_CASE() checks whether one of the termination rules can be applied. BCP() returns either value SUCCESS, FAILURE, or UNSOLVED. If the satisfiability problem cannot be solved by the BCP procedure, the (new) conjunctive normal form χ has to be split with respect to some literal L which is determined by some *branching rule*. Note that, $\chi \cdot L$ is satisfiable only if value 1 is assigned to literal L. Thus, $\chi \cdot L$ is satisfiable if and only if the cofactor of χ with respect to $L = 1$ is satisfiable. Thus the splitting phase of the above implementation realizes the splitting rule of DLL.

1.2.3 Branching rules

In the splitting phase, a literal L which occurs in χ is chosen and DLL generates two sub-problems: the first one in which value 1 is assigned to L and the second one in which L is set to value 0. It has been experimentally shown that the choice of the literal which is used for splitting makes an impact on the efficiency of DLL. The *branching rule* a SAT algorithm uses is one of the dominant factors in its success.

Following the satisfaction hypothesis [61] "Other things being equal, a branching rule performs better when it creates simpler subproblems", good branching heuristics are those which choose a literal L such that BCP($\chi \cdot L$) and BCP($\chi \cdot L'$) are as simple as possible. Note that BCP($\chi \cdot L$) and BCP($\chi \cdot L'$) extend the current (partially defined) assignment to the variables by following the logical consequences of the assignment $L = 1$ and $L = 0$, respectively.

In the following, let us present the branching rules mostly applied in practice [49, 59, 61, 69, 90, 97, 148]. All of them give preference to variables or literals which occur in small clauses. By this means, the conjunctive normal form χ is quickly transformed into a form which contains one-literal clauses such that BCP can be applied and conflicts can be found. In Section 1.2.4 we will see that detected conflicts help prune the search space.

MOM's Heustistic. *Maximum Occurrences on clauses of Minimum sizes* (MOM) is a branching rule which gives preference to variables that occur frequently as positive as well as negative literal in the shortest clauses of χ. More formally, let

$$f_k(L) = |\{c;\ c \in \chi, |c| = k, \text{ and } L \in c\}|$$

be the number of occurrences of literal L in clauses which consists of k literals and κ be the length of a shortest clause of χ, then $f_\kappa(L)$ denotes the number of occurrences of literal L in the shortest (unresolved) clauses of χ. Now, MOM's heuristic selects a literal L such that $f_\kappa(L) + f_\kappa(L')$ is maximal. In order to obtain subproblems of approximately equal sizes, the minimum of these two quantities $f_\kappa(L)$ and $f_\kappa(L')$ should be maximal. These criteria can be modelled by

$$\max_{\text{literal } L} \left\{(f_\kappa(L) + f_\kappa(L')) \cdot 2^s + f_\kappa(L) \cdot f_\kappa(L')\right\}$$

for some adequate constant s. (Cf. [49, 94]).

Jeroslow-Wang rules. There are two Jeroslow-Wang [69] branching rules. Both rules give preference to literals that occur often in small clauses of χ as well, but in a more sophisticated way than MOM's heuristic does. They are based on the function JW which associates a literal L with the weight

$$JW(L) = \sum_{c \in \chi,\ L \in c} 2^{-|c|}.$$

The *one-sided JW branching rule* selects the literal L with the largest value $JW(L)$. The *two-sided JW branching rule* selects a variable x_i such that the combined value $JW(x_i) + JW(x_i{}')$ is maximal. In the algorithm shown in Figure 5.3, literal L is set to x_i, if $JW(x_i) \geq JW(x_i{}')$ holds (and the two-sided JW branching rule is applied).

1.2.4 Pruning the search space by using detected conflicts

Given an initial conjunctive normal form χ, one can attempt to augment it with additional clauses to increase the deductive power during the search process. This technique is called *clause recording*, in literature (cf. [80]). Let us justify the statement just made by an example.

Let χ be the conjunctive normal form defined by

$$\begin{aligned}
&(x_9{}' + x_{14}) \cdot (x_{10} + x_{15}{}') \cdot (x_{11}{}' + x_{16}) \cdot (x_1{}' + x_2 + x_9) \\
&\cdot (x_1 + x_{11}) \cdot (x_1{}' + x_3 + x_9 + x_{10}{}' + x_{11}) \cdot (x_2{}' + x_3{}' + x_4) \\
&\cdot (x_4{}' + x_5 + x_6{}') \cdot (x_4{}' + x_6) \cdot (x_5{}' + x_6{}') \\
&\cdot (x_1 + x_7 + x_{12}{}') \cdot \chi_1
\end{aligned}$$

where χ_1 is some adequate conjunctive normal form. Let us assume that BCP is not applicable to formula χ because of subformula χ_1 and that literal $x_9{}'$ is chosen by the branching rule, i.e., SATISFY($\chi \cdot x_9{}'$) is called. Applying BCP to $\chi \cdot x_9{}'$ results in

$$\begin{aligned}
&(x_{10} + x_{15}{}') \cdot (x_{11}{}' + x_{16}) \cdot (x_1{}' + x_2) \cdot (x_1 + x_{11}) \\
&\cdot (x_1{}' + x_3 + x_{10}{}' + x_{11}) \cdot (x_2{}' + x_3{}' + x_4) \cdot (x_4{}' + x_5 + x_6{}') \\
&\cdot (x_4{}' + x_6) \cdot (x_5{}' + x_6{}') \cdot (x_1 + x_7 + x_{12}{}') \cdot \chi_2.
\end{aligned}$$

Once again, we assume that the algorithm has to branch and that literal x_{10} is chosen by the branching rule this time. Applying BCP to the subproblem BCP($\chi \cdot x_9{}'$) $\cdot\, x_{10}$ results in

$$\begin{aligned}
&(x_{11}{}' + x_{16}) \cdot (x_1{}' + x_2) \cdot (x_1 + x_{11}) \cdot (x_1{}' + x_3 + x_{11}) \\
&\cdot (x_2{}' + x_3{}' + x_4) \cdot (x_4{}' + x_5 + x_6{}') \cdot (x_4{}' + x_6) \\
&\cdot (x_5{}' + x_6{}') \cdot (x_1 + x_7 + x_{12}{}') \cdot \chi_3.
\end{aligned}$$

In the next iteration, literal $x_{11}{}'$ may be chosen to split the problem and we obtain

$$\begin{aligned}
&(x_1{}' + x_2) \cdot (x_1) \cdot (x_1{}' + x_3) \cdot (x_2{}' + x_3{}' + x_4) \cdot (x_4{}' + x_5 + x_6{}') \\
&\cdot (x_4{}' + x_6) \cdot (x_5{}' + x_6{}') \cdot (x_1 + x_7 + x_{12}{}') \cdot \chi_4.
\end{aligned}$$

Now, BCP assigns value 1 to variable x_1 because there is a clause which only consists of literal x_1. This assignment reduces the above formula to the conjunctive normal form

$$(x_2) \cdot (x_3) \cdot (x_2' + x_3' + x_4) \cdot (x_4' + x_5 + x_6') \\ \cdot (x_4' + x_6) \cdot (x_5' + x_6') \cdot \chi_5.$$

Once again, there are one-literal clauses and BCP assigns value 1 to the literals x_2, x_3, and x_4 in the further iterations which results in

$$(x_5 + x_6') \cdot (x_6) \cdot (x_5' + x_6') \cdot \chi_8.$$

Because of the one-literal clause $\{x_6\}$, BCP transforms this conjunctive normal form into

$$(x_5) \cdot (x_5') \cdot \chi_9,$$

which results in a conflict in the next iteration of BCP. This conflict is due to the decisions made by the branching rule, namely the assignment of value 0 to variable x_9, value 1 to variable x_{10}, and value 0 to variable x_{11}. Because of the conflict, any satisfying assignment of χ must satisfy clause $x_9 + x_{10}' + x_{11}$. Thus, χ can be augmented by this new clause which is called *conflict induced clause*. The conflict induced clause averts future occurrences of the same conflict as whenever two of the three literals are deleted from this clause by DLL, the one-literal clause generated in this way takes care that we are not led to the same conflict.

Different conflict analysis procedures which compute conflict induced clauses have been proposed in literature. We refer to [95, 96, 97, 112, 148, 149]. We will describe the underlying ideas of conflict analysis procedures in the next section.

1.2.5 Conflict analysis procedure

In order to obtain efficient conflict analysis procedures, each branching decision made by SPLIT() and each implication drawn by BCP() is captured by an *implication graph* [97] which is a directed graph defined as follows:

- Each node of the implication graph is either an *assignment node* or a *conflict node*. There is at most one conflict node.

- Each *assignment node* corresponds to an assignment $b \in \{0, 1\}$ of a variable x_i of the initial conjunctive normal form under consideration. In DLL, variable assignments are carried out either by the splitting phase or the BCP phase. As variable assignments made by the BCP phase are logical implications, the corresponding nodes of the implication graph are called *induced nodes*. The nodes which correspond to variable assignments carried out by the branching rule are called *decision nodes*. Each node is tagged with this information. Note that for the sake of convenience we will often call a node by its corresponding variable assignment.

- An induced node v has predecessors. The predecessors are those variable assignments which directly imply the variable assignment node v stands for. More formally, let $c = \{L_1 \ldots, L_k\}$ be a clause of the initial conjunctive normal form under consideration . If the SAT solver has assigned value 0 to each of the literals $L_1, \ldots L_{k-1}$, then literal L_k has to be assigned value 1 in order to satisfy clause c (see the one-literal clause rule). Thus, the induced node $L_k = 1$ and $k-1$ edges $e_1, \ldots, e_{k-1}$ are inserted into the implication graph by the BCP phase. Edge e_j leaves node $L_j = 0$ and enters the induced node $L_k = 1$.

- The conflict node has predecessors as well. The predecessors of a conflict node are those nodes whose variable assignments are directly responsible for the conflict.

- A decision node has no incident predecessor.

The implication graph is used for conflict analysis *and* performing implications. As performing implications is a very time-consuming task in formal equivalence checking, it is very important to use an efficient and suited data structure. For more details, we refer to [137].

Figure 5.4a shows the implication graphs of the first five steps carried out in the example shown in Section 1.2.4. For illustration, we have enumerated the clauses of the initial conjunctive normal form χ:

$$1 : ({x_9}' + x_{14})$$

$$2 : (x_{10} + {x_{15}}')$$

$$3 : ({x_{11}}' + x_{16})$$

$$4 : ({x_1}' + x_2 + x_9)$$

$$5 : (x_1 + x_{11})$$

$$6 : ({x_1}' + x_3 + x_9 + {x_{10}}' + x_{11})$$

$$7 : ({x_2}' + {x_3}' + x_4)$$

$$8 : ({x_4}' + x_5 + {x_6}')$$

$$9 : ({x_4}' + x_6)$$

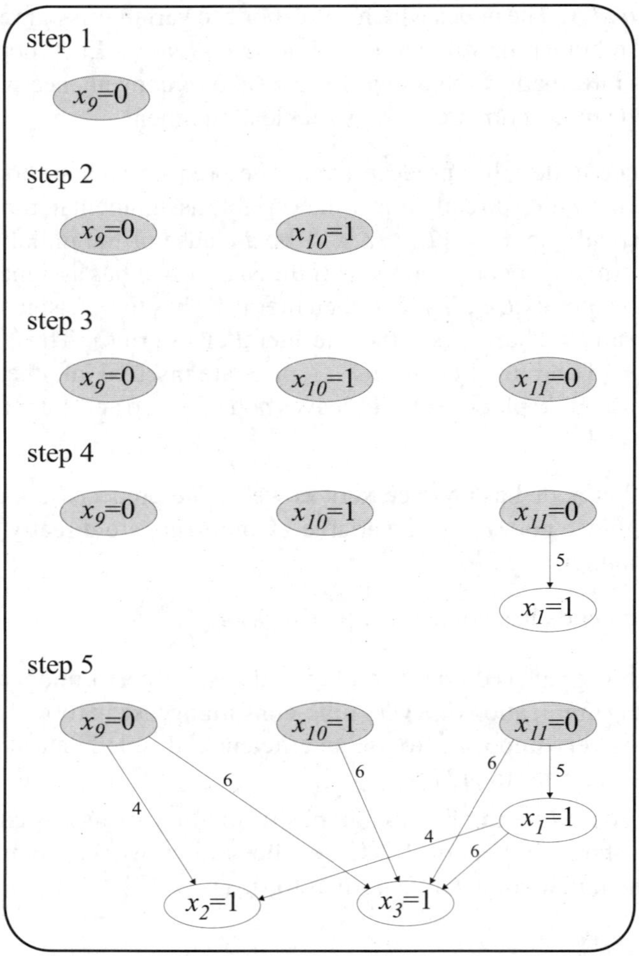

Figure 5.4a. First five implication graphs of the example from above.

10 : $(x_5' + x_6')$

11 : $(x_1 + x_7 + x_{12}')$

12 : ...

Decision nodes are marked by hatching. Edges are tagged with the clause which is responsible for the implication. For example, in Step 5 the sixth clause

$$(x_1' + x_3 + x_9 + x_{10}' + x_{11})$$

is responsible for the assignment of value 1 to variable x_3 as all its literals but x_3 are not satisfied by the variable assignment made by the time. Thus, there are edges from the nodes $x_1 = 1$, $x_9 = 0$, $x_{10} = 1$, and $x_{11} = 0$ to the induced node $x_3 = 1$. The final implication graph after having detected the conflict is shown in Figure 5.4b.

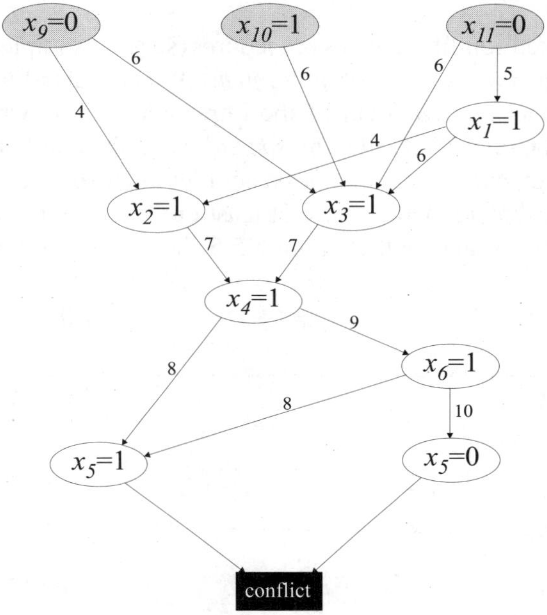

Figure 5.4b. Final implication graph.

Given an implication graph with a conflict node, the variable assignments carried out by the branching rule which are responsible for the conflict node can be gathered by inverting all the edges of the implication graph and exploring the graph starting at the conflict node. An exploration path is aborted if a decision node is reached. More formally, given a node v of the implication graph, the set of decision nodes which are responsible for the corresponding variable assignment is given by

$$causes(v) = \begin{cases} v, & \text{if } v \text{ is a decision node} \\ \bigcup_{w \in pred(v)} causes(w), & \text{if } v \text{ is an induced node} \end{cases}$$

where

$$pred(v) = \{w;\ (w, v) \text{ is an edge of the implication graph}\}$$

is the set of the direct predecessors of v in the current implication graph (cf. [97]).
In our example the assignments $x_9 = 0$, $x_{10} = 1$, and $x_{11} = 0$ are responsible for the conflict, i.e., if the formula $x_9' \cdot x_{10} \cdot x_{11}'$ evaluates to 1, then the above conflict appears. Negating the formula results in the clause $(x_9 + x_{10}' + x_{11})$. This clause has to be satisfied for every variable assignment which satisfies the initial conjunctive normal clause χ.

More sophisticated conflict analysis procedures (see for example GRASP (*Generic seaRch Algorithm for the Satisfiable Problem*) [97] and CHAFF [112]) take advantage of the fact that *each* cut of the implication graph which has all the decision nodes on one side (called *reason side*) and the conflicting node at the other side (called *conflict side*) defines a conflict induced clause [149]. All the nodes on the reason side which have at least one edge to the conflict side are part of the reason for the conflict. Figure 5.5 shows three different cuts which

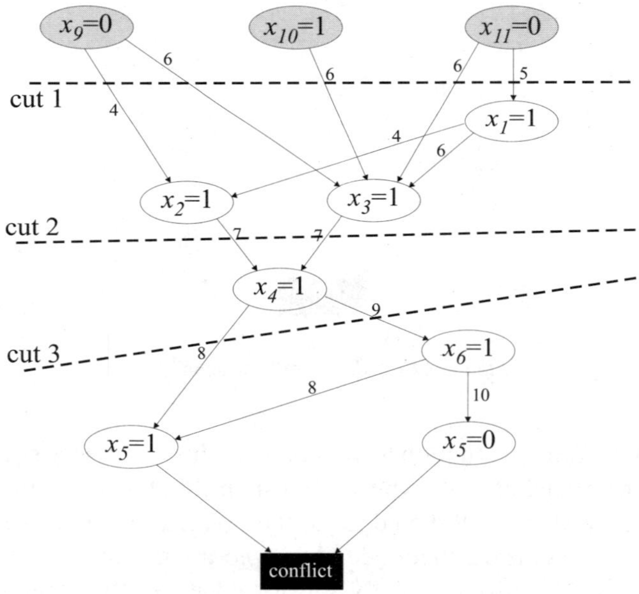

Figure 5.5. Some cuts of the implication graph.

imply the conflict induced clauses

$$x_9 + x_{10}' + x_{11},$$

$$x_2' + x_3',$$

and

$$x_4{}',$$

respectively. This observation leads to the idea of generating stronger conflict induced clauses, i.e., clauses which contain fewer literals. This can be done by a more careful analysis of the structure of the implication graph. For example, in [97] so called *unique implication points* play a decisive role in this context. A unique implication point with respect to a detected conflict represents a variable assignment which, by itself, can trigger sequences of implied assignments that yield the conflict.

DEFINITION 5.3 (UNIQUE IMPLICATION POINT) *A node v of the implication graph is said to dominate another node w of the implication graph, if and only if any path from the latest decision node to node w needs to go through node v. A node which dominates the conflict node of the implication graph is called unique implication point.*

For illustration, consider the implication graph shown in Figure 5.4b. Node $x_4 = 1$ is a unique implication point. Indeed, the induced variable assignment $x_4 = 1$ yields by itself the identified conflict. The more general case is shown in Figure 5.6. Let us assume that node $x_7 = a$ $(a \in \{0, 1\})$ is the latest decision node. Then, node $x_8 = b$ $(b \in \{0, 1\})$ dominates the conflict node as any path from node $x_7 = a$ to the conflict node goes through node $x_8 = b$. In GRASP [97], two clauses are added to the clause database if a unique implication point is found. The first clause corresponds to the cut where the variable assignments induced by the unique implication point are put on the conflict side and the remaining nodes are put on the reason side. In the implication graph of Figure 5.6, this clause is given by

$$x_1 + x_2{}' + x_3 + x_4{}' + x_8^{b'}$$

where x_8^0 and x_8^1 denote $x_8{}'$ and x_8, respectively. The second clause added by GRASP describes the reconvergence between the latest decision node and the unique implication point. That is, clause

$$x_8^b + x_3 + x_4{}' + x_7^{a'} + x_5 + x_6{}',$$

which describes the logical implication

$$(x_3{}' \cdot x_4 \cdot x_7^a \cdot x_5{}' \cdot x_6) \Rightarrow x_8^b,$$

is added in our example. Note, that, in some sense, the two clauses completely specify the current conflict.

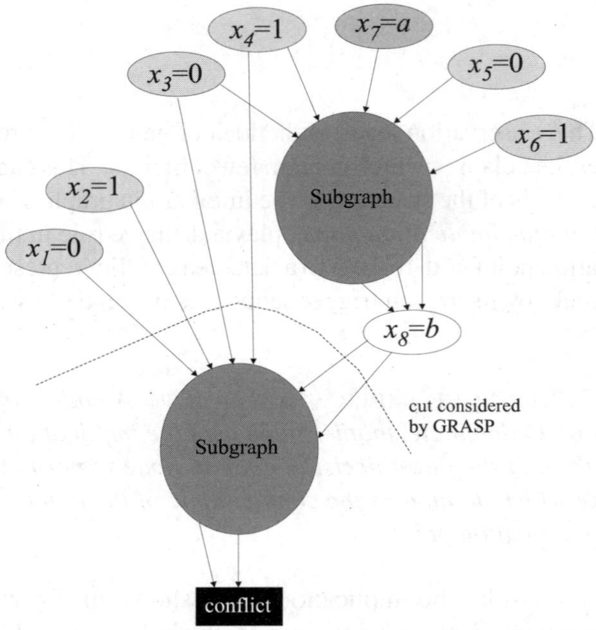

Figure 5.6. Unique implication point.

1.3 Using BDDs to speed up SAT methods

In [55] and [123], Boolean satisfiability methods have been integrated with BDDs to reduce both the problem size and the number of backtracks for the satisfiability problem. We review both approaches in the next two subsections.

1.3.1 The approach of Gupta and Ashar

In the approach of [55], the miter is partitioned such that the output of the miter is on one side and the primary inputs of the miter are on the other side. A single BDD is used to capture the *fanout part*, that is the partition of the miter near its output. SAT clauses capture the *fanin part*. For illustration, see Figure 5.7 which shows a miter partitioned into a fanin part and a fanout part. The ROBDD which describes the fanout part is shown in Figure 5.7 as well. The conjunctive normal form $\chi(\mathcal{N}_{fanin})$ which describes the consistent signal line

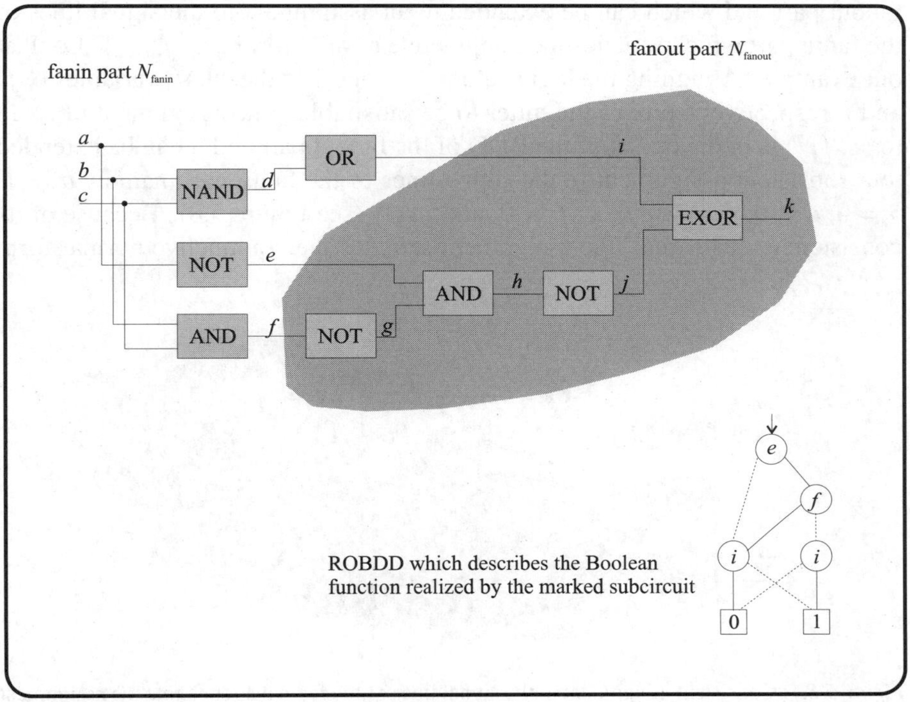

Figure 5.7. Illustration of the approach of Gupta and Ashar [55].

assignments of the fanin part of the circuit is given by

$$\begin{aligned} \chi(\mathcal{N}_{fanin}) \quad = \quad & (d+b)\cdot(d+c)\cdot(d'+b'+c') \\ & \cdot\ (e'+a')\cdot(e+a) \\ & \cdot\ (f'+b)\cdot(f'+c)\cdot(f+b'+c') \\ & \cdot\ (i+a')\cdot(i+d')\cdot(i'+a+d). \end{aligned}$$

The BDD and the conjunctive normal form share variables, namely those which correspond to signal lines crossed by the boundary between the fanin part and the fanout part (which we call *cut* in the following as well). These variables are called *cutset variables*. In the above example, the set of the cutset variables consists of the signal lines i, e, and f.

Now, remember that the objective is to check whether there is an assignment to the primary inputs of the miter such that its output line evaluates to value 1. It is straightforward to see that the miter is satisfiable if and only if there is an assignment to the cutset variables which is in the on-set of the BDD of the fanout part and which can be extended to an assignment to the signal lines of the fanin part which satisfies the conjunctive normal form $\chi(\mathcal{N}_{fanin})$. Look at our example. Assigning the logic values 1, 0, and 1 to the cutset variables e, f, and i, respectively, proves the miter to be satisfiable. The assignment $(1, 0, 1)$ to (e, f, i) is in the on-set of the BDD of the fanout part and it can be extended to a consistent assignment to the signal lines of the fanin part, namely $a = 0$, $b = 0$, $c = 0$, $d = 1$, $e = 1$, $f = 0$, and $i = 1$ (see Figure 5.8). Because of its consistency, this signal line assignment satisfies the conjunctive normal form $\chi(\mathcal{N}_{fanin})$.

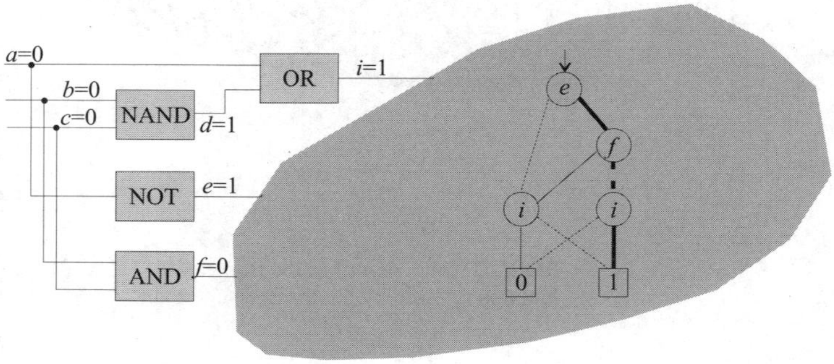

Figure 5.8. Consistent assignment to the signal lines of the fanin part such that the assignment of the cutset variables is in the on-set of the BDD of the fanout part.

A naive implementation of this approach is to enumerate each path which starts at the root of the BDD of the fanout part and ends at terminal node $\mathbf{1}$. For each such path, the (partial) assignment to the cutset variables can be taken to augment the conjunctive normal forms by the corresponding one-literal clauses. In the above example, there are three such paths which correspond to the partial variable assignments

- $e = 0$ and $i = 0$,
- $e = 1$, $f = 1$, and $i = 0$, and
- $e = 1$, $f = 0$, and $i = 1$,

respectively. Thus, the three SAT Problems

- $\chi(\mathcal{N}_{fanin}) \cdot (e') \cdot (i')$,
- $\chi(\mathcal{N}_{fanin}) \cdot (e) \cdot (f) \cdot (i')$, and
- $\chi(\mathcal{N}_{fanin}) \cdot (e) \cdot (f') \cdot (i)$

have to be solved, independently of each other. If any of these problems has a solution, the miter is satisfiable and the two circuits to be compared are not functionally equivalent.

A more sophisticated way has been proposed by Gupta and Ashar [55]. In their approach, the SAT engine is applied to the conjunctive normal form $\chi(\mathcal{N}_{fanin})$ of the fanin part and proceeds as it is described in the above sections. However, whenever a cutset variable is being bound in the SAT procedure, they check whether the partial variable assignment obtained so far has a non-null intersection with the on-set of the BDD, i.e., whether the partial assignment to the cutset variables can be extended to a full assignment to the cutset variables which is in the on-set of the BDD. If it does not, each extension of the current partial assignment to the cutset variables can be ruled out. This *Early Bounding* approach essentially prunes off large subspaces of the decision tree.

1.3.2 The approach of Reda and Salem

In the approach of [123], the miter is also partitioned such that the primary output of the miter is on one side and the primary inputs of the miter are on the other side. However, the conjunctive normal form is built for the fanout part and the ROBDD F_s of every cutset variable s is built. Figure 5.9 illustrates the idea. It shows the ROBDDs of the cutset variables as well. The conjunctive normal form $\chi(\mathcal{N}_{fanout})$ which describes the consistent signal line assignments of the fanout part of the circuit is given by

$$
\begin{aligned}
\chi(\mathcal{N}_{fanout}) \quad = \quad & (g' + f') \cdot (g + f) \\
& \cdot\ (h' + e) \cdot (h' + g) \cdot (h + e' + g') \\
& \cdot\ (j' + h') \cdot (j + h) \\
& \cdot\ (k' + i + j) \cdot (k' + i' + j') \\
& \cdot\ (k + i' + j) \cdot (k + i + j').
\end{aligned}
$$

In order to check the consistency of the variable assignments made by the SAT engine applied to $\chi(\mathcal{N}_{fanout}) \cdot k$, a ROBDD B is used to store the characteristic function of the set of the assignments to the primary miter inputs. These inputs imply the variable assignments to the cutset variables already bound by the SAT engine. The ROBDD B is initialized to the constant BDD $\mathbf{1}$. Now, whenever a cutset variable v is assigned a value $\alpha_v \in \{0, 1\}$ by the satisfiable engine, ROBDD B is updated. It is intersected with either F_v if $\alpha_v = 1$ or $NOT(F_v)$ if $\alpha_v = 0$. If at any time ROBDD B reduces to $\mathbf{0}$, a contradiction has been found. The SAT engine has to backtrack.

For illustration, consider the circuit shown in Figure 5.9. Assume that the SAT engine decided to assign value 0 to variable i and value 1 to variable j. The latter decision directly implies the variable assignment $h = 0$. After these

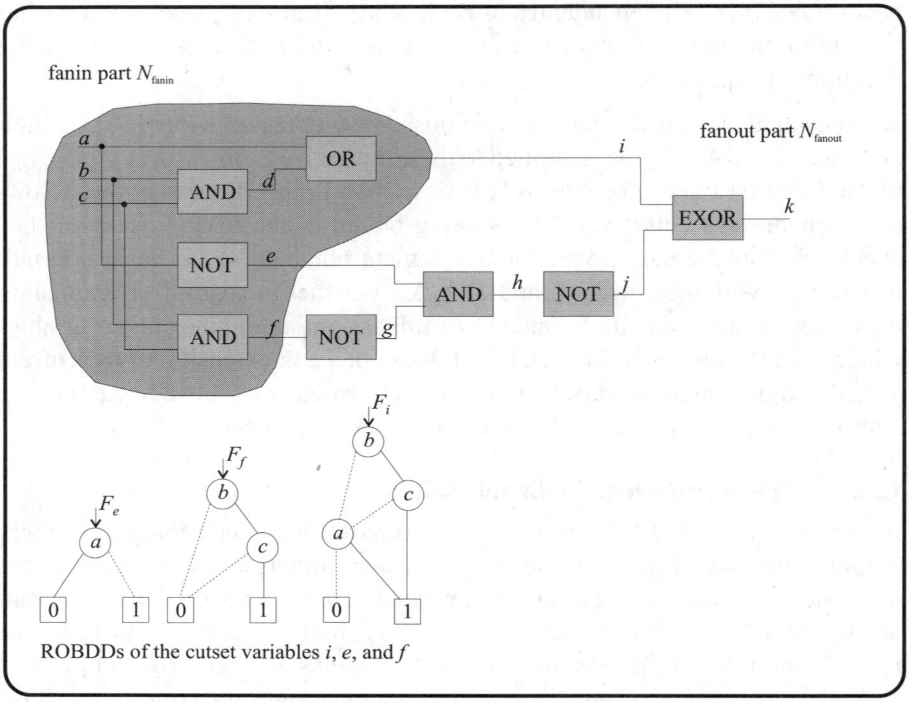

Figure 5.9. Illustration of the approach of Reda and Salem [123].

variable assignments, the current conjunctive normal formal of the fanout part is given by

$$(g' + f') \cdot (g + f) \cdot (e' + g').$$

At this time, ROBDD B is equal to the ROBDD $NOT(F_i)$ as variable i is the only cutset variable which is bound. Now assume that the SAT engine decides to assign value 0 to g. This directly induces that value 1 is assigned to cutset variable f. The conjunctive normal form above is satisfied. However, the intersection of the current ROBDD B and ROBDD F_f is $\mathbf{0}$. Thus, there is no assignment to the miter inputs which justifies the current variable assignment of the cutset variables i and f. A conflict has been detected and the SAT engine has to backtrack.

2. ATPG based equivalence checking

ATPG (*automated test pattern generation*) based equivalence checking is closely related to SAT based equivalence checking. The most important difference between SAT and ATPG results from the fact that SAT-algorithms operate on conjunctive normal forms while ATPG-algorithms generally operate on Boolean networks. While research around SAT started in the context of automated theorem proving and was regarded as a subfield of artificial intelligence, research in ATPG was primarily driven by circuit testing. For a detailed comparison of SAT and ATPG we refer to [80].

Section 2.1 and Section 2.2 partly follow [1].

2.1 Introduction

In the following we give a short introduction to circuit testing, and show the interrelation of circuit testing and combinational equivalence checking.

2.1.1 Circuit testing

Let $\mathcal{N}$ be a Boolean network and $\phi(\mathcal{N}) \in \mathcal{B}_{n,m}$ be the Boolean function represented by $\mathcal{N}$. The presence of a fabrication fault ρ transforms $\mathcal{N}$ into a new circuit $\mathcal{N}_\rho$. Let us assume that $\mathcal{N}_\rho$ is a Boolean network as well, and that its behavior is described by the Boolean function $\phi(\mathcal{N}_\rho) \in \mathcal{B}_{n,m}$. The problem of testing for fault ρ is to find an input vector $\alpha \in \{0,1\}^n$ such that

$$\phi(\mathcal{N})(\alpha) \neq \phi(\mathcal{N}_\rho)(\alpha)$$

holds. Such a vector α is called *test vector* or *distinguishing vector* for fault ρ. If $\mathcal{N}$ is a hardware implementation of a single-output Boolean function, then a vector α is a test vector for some fault ρ if and only if

$$\phi(\mathcal{N})(\alpha) \oplus \phi(\mathcal{N}_\rho)(\alpha) = 1.$$

2.1.2 The stuck-at-fault model

Physical faults are partly modelled by logical faults. This approach reduces the complexity of testing as many different physical faults can be modelled by the same logical fault. Moreover, the problem of fault analysis becomes a logical rather than a physical problem.

In many technologies, a short between ground or power and a signal line or an open on a unidirectional signal line can make the signal remain at a fixed voltage level [33]. This behavior is modelled by the *stuck-at-fault model*. A *stuck-at-fault* is a logical fault which consists of a signal being stucked at a fixed logic value $b \in \{0,1\}$. It is denoted by s-a-b.

For testing a signal e for s-a-b, a vector has to be found that activates the fault, i.e., assigns value b' to signal e in the faultless circuit. This phase is usually called *justification process* or *activation process*. Thus, different values occur at the site of the fault in the faultless and in the faulty circuit. This difference has to be propagated to a primary output.

2.1.3 Testing and equivalence checking

Let us look at the miter shown in Figure 5.1. The Boolean networks F and G are functionally equivalent if and only if there is no assignment to the primary inputs of the miter such that the output line evaluates to 1. To test the output for s-a-0, a test vector has to be applied that activates the fault, i.e., satisfies the miter. Thus, in some sense, combinational equivalence checking is a special case of testing for stuck-at-faults. Only the activation phase has to be carried out, the propagation phase can be omitted.

2.2 Basic methods for ATPG based equivalence checking

As just mentioned, we only have to concentrate on the activation phase. Consider the miter of Figure 5.2, once again. In order to justify value 1 at output k, either signal line i or signal line j has to be assigned value 1. Thus, the search for a solution involves a decision process. Let us first try to justify $i = 1$ and $j = 0$. For illustration, please see Figure 5.10. By backward implication, the

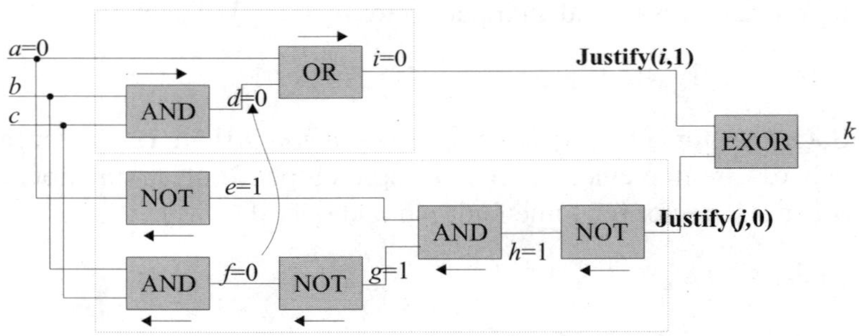

Figure 5.10. Justifying $i = 1$ and $j = 0$

signal assignment $j = 0$ induces assignment $h = 1$. Iteratively, it implies the assignments $e = 1$, $g = 1$, $a = 0$, and $f = 0$ by backward implication. To justify $f = 0$, signal line b or signal line c has to be assigned value 0. Both alternatives directly induce that value 0 has to be assigned to signal line d. As both, signal line a and signal line d are assigned value 0, signal line i has to be 0 as well. However, this conflicts with the fact, that we wanted to justify $i = 1$. Thus, we have to backtrack and try to justify $i = 0$ and $j = 1$. Since this justification process results in a conflict also, there is no assignment to the primary inputs of the miter which activates s-a-0 at output k. Thus, output k cannot be tested for s-a-0 and the two circuits are functionally equivalent.

2.2.1 Justification procedure

For the justification procedure, in addition to the values 0 and 1 a third logic value X which describes the *unknown* value is required. In a preprocessing step, value X is assigned to each signal line but the primary output signal of the miter which is assigned value 1.

JUSTIFY()
begin

if IMPLICATION()=FAILURE **then** return FAILURE **fi**;

if *J-frontier* is the empty set **then** return SUCCESS **fi**;

select a gate G from *J-frontier*;

repeat

select one untried (partial) assignment of the inputs of G,
which justifies the assignment at the output of G, and
update *J-frontier*;

if JUSTIFY()=SUCCESS **then** return SUCCESS **fi**;

restore *J-frontier*;

until no untried assignment of the inputs of G exists;

return FAILURE;

end

Figure 5.11. Justification procedure

During the justification process, we have to keep track of the currently unjustified output signals of gates, i.e., output signals of gates which have been assigned a value of 0 or 1, but where this value is not yet implied by the logic values at the gate inputs. The set is called *J-frontier*. Thus, at the beginning of the justification process *J-frontier* consists of the output signal line of the miter only.

Figure 5.11 outlines the justification procedure. Procedure IMPLICATION() computes all signal line assignments which can uniquely be determined by implication, maintains the *J-frontier*, and checks for consistency. If the consistency check fails, the procedure reports FAILURE. First of all, *local implications* are applied by procedure IMPLICATION(). Local implications are implications which propagate values over one basic gate. Exemplarily, the local implications which correspond to the AND-gate and the EXOR-gate are shown in Table 5.2a and Table 5.2b, respectively. z denotes the output signal line of the basic gate. The input signal lines are denoted by x_1 and x_2.

$type(v)$	**current assignments**			**implied assignments**		
	z	x_1	x_2	z	x_1	x_2
$\cdot$	1	X	X	1	1	1
$\cdot$	1	1	X	1	1	1
$\cdot$	1	X	1	1	1	1
$\cdot$	0	1	X	0	1	0
$\cdot$	0	X	1	0	0	1
$\cdot$	X	1	1	1	1	1
$\cdot$	X	0	0	0	0	0
$\cdot$	X	0	1	0	0	1
$\cdot$	X	1	0	0	1	0

Table 5.2a. The local implications of the AND-gate

$type(v)$	**current assignments**			**implied assignments**		
	z	x_1	x_2	z	x_1	x_2
$\oplus$	1	0	X	1	0	1
$\oplus$	1	X	1	1	0	1
$\oplus$	1	1	X	1	1	0
$\oplus$	1	X	0	1	1	0
$\oplus$	0	X	1	0	1	1
$\oplus$	0	1	X	0	1	1
$\oplus$	0	X	0	0	0	0
$\oplus$	0	0	X	0	0	0
$\oplus$	X	0	0	0	0	0
$\oplus$	X	1	1	0	1	1
$\oplus$	X	0	1	1	0	1
$\oplus$	X	1	0	1	1	0

Table 5.2b. The local implications of the EXOR-gate

The efficiency of the justification process strongly depends on the power of the local and *global implications* used. Global implications have to be learned. We are going to present learning strategies in Section 2.2.2. These results are mainly due to Kunz [82].
After application of the implication rules, the justification process checks whether there are unjustified signal assignments. If the J-frontier is empty, then all the signal assignments made are justified by the signal assignments to the primary inputs of the miter. Otherwise, a gate G whose output is unjustified is chosen. At this point, (usually) there are several alternative ways to solve the local justification problem. For example, assume that G is an AND-gate whose input signal lines are assigned value X and whose output signal is assigned value 0. Then, the justification process has to assign value 0 to one of the input signal lines of G. The algorithm selects one of the alternative ways and tries to solve the problem. This process continues until a solution is found or all possible choices have failed. More details on this basic backtracking algorithm can be found in [1]

2.2.2 Learning global implications

As already mentioned, the efficiency of the justification process strongly depends on the available implications rules. In addition to local implication rules, global implication rules which involve larger areas of the circuit and reconvergent fanouts are necessary to obtain a powerful justification engine [133, 134, 135]. Figure 5.12 shows the circuit often used for illustration of global implications [1, 82]. Obviously, the assignment $c = 1$ at the output line

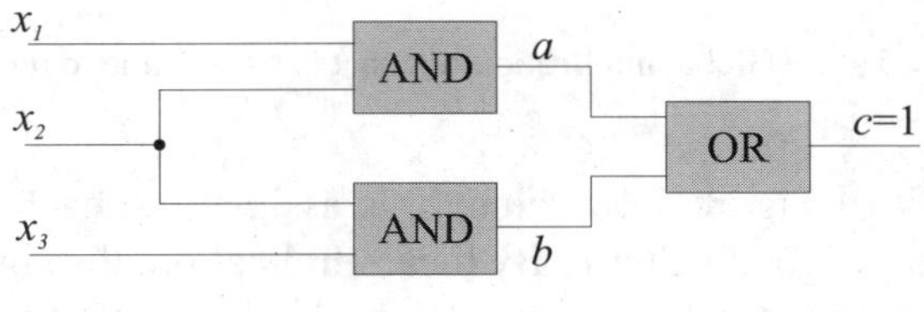

Figure 5.12. Global implication rule.

of the OR-Gate cannot be justified by local implications. At first glance, we have to decide whether we try to justify $c = 1$ by setting either $a = 1$ or $b = 1$, at first. However, a closer look reveals that x_2 has to be 1, independently of the chosen alternative. This can easily be ´´learned´´ by assigning value X to all signal lines but x_2 which is assigned value 0. By local implications, it can be concluded that signal lines a and b are assigned value 0, too. Thus, the assignment $x_2 = 0$ implies the assignment $c = 0$. Thus, we have obtained the implication

$$x_2 = 0 \Rightarrow c = 0.$$

By means of contraposition, the implication transforms into

$$c = 1 \Rightarrow x_2 = 1$$

which is of larger interest, since it cannot be obtained with local implications only.

Schulz et.al. [134] proposed to conduct learning in a preprocessing phase, i.e., before the justification process is invoked. This approach which is called *static learning* tries to identify logic relations between signals of the circuit under consideration. They assign the values 0 and 1 after each other to every signal in the circuit and examine their logical consequences by applying local implication rules and global implication rules stored already before. These implication rules are transformed by contraposition. A learning criterion checks whether the implications derived by contraposition can be directly derived by local implications as well. In this case, it is not worthwhile to record the implication. For illustration, consider the circuits shown in Figure 5.13 and 5.14. Assume

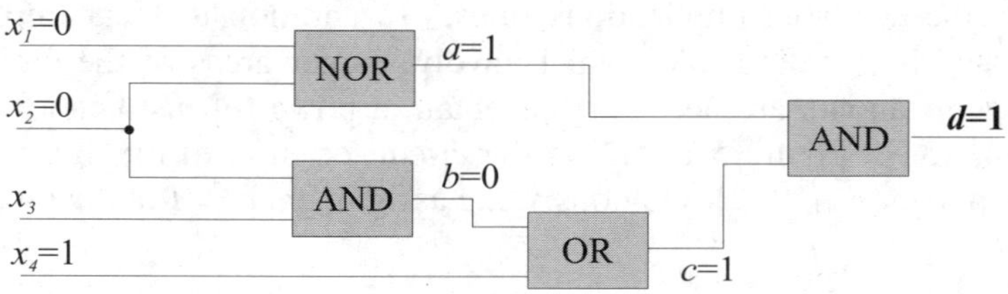

Figure 5.13. Global implication rule that is worthwhile to record.

that in the circuit of Figure 5.13 value X is assigned to each signal line but signal d which is assigned value 1. By local implication, the signal lines a and c have to been assigned value 1, too. Because of $a = 1$, the primary input signals x_1 and x_2 have to be set to value 0 which directly induces $b = 0$. As signal lines b and c are assigned value 0 and 1, respectively, the primary input signal x_4 has to be set to value 1 in order to justify $d = 1$. Thus, by the iterated local implication process, the assignment of value 1 to signal line d directly implies signal assignment $x_4 = 1$. By means of contraposition, this implication transforms into the implication

$$x_4 = 0 \Rightarrow d = 0.$$

A closer look reveals that this new implication cannot be derived by applying local implications only. Thus, it is worthwhile to learn and record it.

Now consider the circuit shown in Figure 5.14 and assume that value X is assigned to each signal line but signal line a which is assigned value 1. The

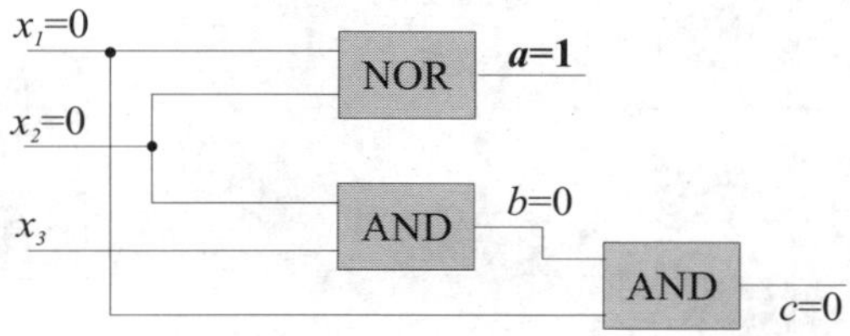

Figure 5.14. Implication rule that is not worthwhile to record.

direct consequences found by local implications are $x_1 = 0$, $x_2 = 0$, $b = 0$, and $c = 0$. Thus, the implication

$$a = 1 \Rightarrow c = 0$$

holds. By contraposition, we obtain the implication

$$c = 1 \Rightarrow a = 0.$$

Obviously, this implication can be directly found by local implications, too. Signal assignment $c = 1$ directly implies $x_1 = 1$ and $x_1 = 1$ directly implies $a = 0$. Thus, it is not worthwhile to learn this "new" implication.

2.2.3 Recursive learning

The techniques just presented do not find all (local and global) implications. Look at the circuit shown in Figure 5.15 and assume that signal line h has been assigned value 0. Obviously, no local implication can be applied. Furthermore, the approach proposed by Schulz et.al. [134] which we have presented in Section 2.2.2 cannot identify a global implication: In order to learn a global implication of the form $h = 0 \Rightarrow \ldots$, it has to find a value assignment to exactly one signal line (but h) which implies $h = 1$. However, such an assignment does not exist because the assignments of signal line e and x_5 are independent from each other.

Nevertheless, there exists a necessary assignment implied by $h = 0$. This implication can easily be found by case differentiation: In order to justify $h = 0$ either $x_5 = 0$ has to hold or $e = 0$ has to hold.

- In the first case, the assignments $g = 1$ and $i = 0$ are directly implied.
- In the second case, the assignments $a = 0$, and $b = 0$ are implied. In order to justify $a = 0$ a new case differentiation has to be performed. In both cases, $x_1 = 0$ and $x_2 = 0$, we observe that $c = 0$ holds. Thus $a = 0$ implies

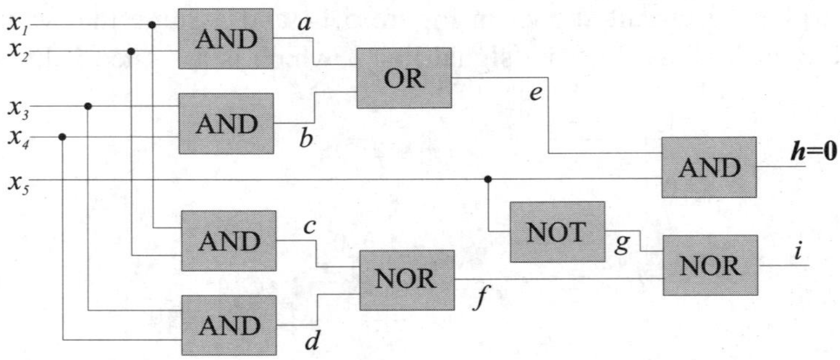

Figure 5.15. Circuit to demonstrate that not all global implications can by found by the techniques presented in Section 2.2.2.

$c = 0$. Analogously, $b = 0$ implies $d = 0$. This implies the assignments $f = 1$ and $i = 0$.

Thus, signal line i has to be assigned value 0 in order to justify $h = 0$. We have found the implication

$$h = 0 \Rightarrow i = 0$$

by case differentiation.

Kunz and Pradhan [81] have developed a method called *recursive learning* which can identify all assignments to signal lines implied from an existing situation of value assignments in the circuit. Readers interested in more details should study [81, 82] where further advanced learning strategies are presented.

2.3 General automatic test pattern generation

In order to fully understand the next chapters, we have to come back to general automatic test pattern generation. As already mentioned on page 121, for testing a signal line e for stuck-at-b with $b \in \{0, 1\}$, a distinguishing vector α has to be found. This assignment α to the primary inputs activates the fault, i.e., assigns value b' to signal e in the faultless circuit. Thus, different values occur at the site of the fault in the faultless and in the faulty circuit. This difference at the fault location has to be propagated to a primary output. Since we only needed to pay attention to the activation phase to understand the verification approach presented in Section 2.2, we have to go into the propagation process, now.

By assigning a value $\alpha \in \{0, 1\}^n$ to the primary inputs of a circuit $\mathcal{N}$ which activates a s-a-b fault ρ, there is at least one signal line s of circuit $\mathcal{N}$ which has different values in the faultless circuit $\mathcal{N}$ and the faulty circuit $\mathcal{N}_\rho$. At least one of these "errors" has to be propagated to a primary output. In order to keep track of the error propagation, Roth [124] proposed to consider composite logic values of the form $\beta_s/\beta_s^{(\rho)}$ where β_s and $\beta_s^{(\rho)}$ are values of the same signal s in $\mathcal{N}$ and $\mathcal{N}_\rho$, respectively. Thus, a 5-valued logic $\{0/0, 1/1, 1/0, 0/1, X/X\}$ is used, where the composite values $0/0$, $1/1$, $1/0$, $0/1$, and X/X are denoted by 0, 1, D, $\overline{D}$, and X [1]. Table 5.3 shows the laws which hold in this logic with respect to the AND-gate, the OR-gate, and the inverter.

OR	0	1	D	$\overline{D}$	X
0	0	1	D	$\overline{D}$	X
1	1	1	1	1	1
D	D	1	D	1	X
$\overline{D}$	$\overline{D}$	1	1	$\overline{D}$	X
X	X	1	X	X	X

AND	0	1	D	$\overline{D}$	X
0	0	0	0	0	0
1	0	1	D	$\overline{D}$	X
D	0	D	D	0	X
$\overline{D}$	0	$\overline{D}$	0	$\overline{D}$	X
X	0	X	X	X	X

NOT	
0	1
1	0
D	$\overline{D}$
$\overline{D}$	D
X	X

Table 5.3. 5-valued logic [1]

Now, in addition to the J-frontier (see page 123) we keep track of the gates whose output line still has been assigned value X and where at least one of its inputs has been assigned value D or value $\overline{D}$. This set is called *D-frontier*. The ATPG methods try to move the J-frontier to the primary inputs and the D-frontier to the primary outputs. This is done by an iterative process in which

- local and global implications are applied, analogously as described for the justification process in Section 2.2, and
- in cases where no implication exists, a gate from the D-frontier or the J-frontier and an assignment of adequate values to the inputs (with value X) of that gate is selected such that the D-frontier or the J-frontier is moved.

The process stops when either at least one primary output of the circuit is assigned value D or value $\overline{D}$ and all assignments are justified by the (partial) assignment to the primary inputs or it detects that the fault injected is not testable. The overall algorithm due to Roth [124] is shown in Figure 5.16. It is an extension of the justification process shown in Figure 5.11. More details can be found in [1].

```
ATPG()
begin
  if IMPLICATION()=FAILURE then return FAILURE fi;

  if no primary output is assigned value D or value D̄
  then

    if D-frontier is the empty set then return FAILURE fi;

    repeat

      select one gate G from D-frontier,
      assign values to the inputs of G such that its output is assigned value D or value D̄,
      and update both sets, J-frontier and D-frontier;

      if ATPG()=SUCCESS then return SUCCESS;

      restore J-frontier and D-frontier and
      undo the assignments made above in this iteration;

    until all gates of D-frontier have been tried;

    return FAILURE;

  fi;

  if J-frontier is the empty set then return SUCCESS fi;

  select a gate G from J-frontier;

  repeat

    select one untried (partial) assignment of the inputs of G,
    which justifies the assignment at the output of G, and
    update J-frontier;

    if ATPG()=SUCCESS return SUCCESS fi;

    restore J-frontier;

  until no untried assignment of the inputs of G exists;

  return FAILURE;

end
```

Figure 5.16. Overall ATPG procedure for Boolean Networks over $\{NOT, \cdot, \bar{\cdot}, +, \overline{+}\}$ [1, 124]

Chapter 6

EXPLOITING SIMILARITIES

Usually, the implementation of a circuit contains a significant number of internal signal lines that have functionally equivalent counterparts in the specification of the circuit [25, 77, 79]. Thus, the corresponding miter (see Figure 5.1 on page 100) is redundant.

The structural similarities can be exploited in order to either simplify the miter so that the corresponding satisfiability problem becomes easier or partition the problem into a set of smaller and simpler combinational equivalence checking problems. In this chapter, we will present both approaches after having reviewed the common basic idea proposed by Berman and Trevillyan [9].

1. The basic idea

One way of exploiting structural similarities during equivalence checking has been proposed in [9]. They proposed a *decomposition based equivalence checker* which proves equivalences between internal signal lines and uses these internal equivalences to check the equivalence of the circuits being compared. The basic idea of their procedure is as follows. Given two signal lines s and t being checked for functional equivalence, they derive a *simultaneous functional decomposition* (h_s, h_t, g) of s and t.

DEFINITION 6.1 (SIMULTANEOUS FUNCTIONAL DECOMPOSITION)
Let $f_s \in \mathcal{B}_n$ and $f_t \in \mathcal{B}_n$ be the Boolean functions which compute the assignments to s and t, respectively. The vector (h_s, h_t, g) with $g \in \mathcal{B}_{n,m}$, $h_s \in \mathcal{B}_m$, and $h_t \in \mathcal{B}_m$ is a simultaneous decomposition of s and t if

$$\begin{aligned} f_s &= h_s \circ g \\ f_t &= h_t \circ g \end{aligned}$$

hold.

Obviously, the equality $h_s = h_t$ would imply the functional equivalence of signal lines s and t.

THEOREM 6.1 *If (h_s, h_t, g) is a simultaneous functional decomposition of s and t, then s and t are functionally equivalent if $h_s = h_t$.*

For illustration, look at the circuits shown in Figure 6.1 [83] which we have already used in Chapter 5 (see page 101). It shows a simultaneous functional decomposition of signal lines i and j. As shown, the assignments to signal

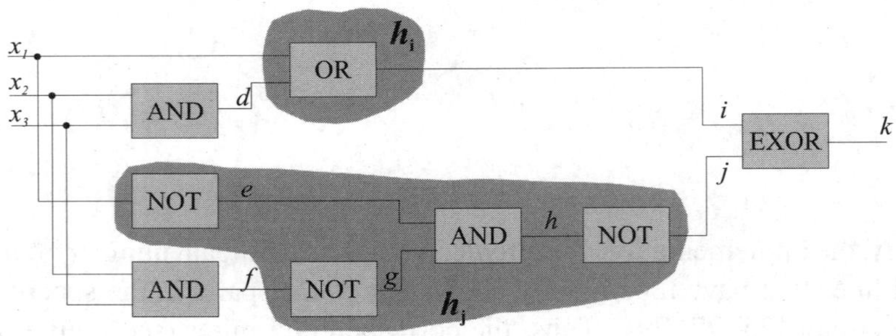

Figure 6.1. Illustration of the decomposition based approach of Berman and Trevillyan.

lines i and j are described by $h_i \circ g$ and $h_j \circ g$ with $g \in \mathcal{B}_{3,2}$ defined by

$$g(x_1, x_2, x_3) = (x_1, x_2 \cdot x_3)$$

and $h_i, h_j \in \mathcal{B}_2$. The signal lines i and j are functionally equivalent since $h_i = h_j$.

Unfortunately, given a simultaneous functional decomposition of two signal lines s and t, the functional equivalence of signal line s and signal line t does not imply that h_s and h_t are equal, in general. As example, look at the simultaneous decomposition given by

$$g : \{0,1\}^2 \rightarrow \{0,1\}^2 \text{ defined by } g(x_1, x_2) = (x_1, x_1 \cdot x_2),$$

$$h_s : \{0,1\}^2 \rightarrow \{0,1\} \text{ defined by } h_s(v, w) = v + w,$$

and

$$h_t : \{0,1\}^2 \rightarrow \{0,1\} \text{ defined by } h_t(v, w) = v.$$

Obviously,

$$\forall (x_1, x_2) \in \{0,1\}^2 : (h_s \circ g)(x_1, x_2) = x_1 = (h_t \circ g)(x_1, x_2)$$

holds, although $h_s \neq h_t$.

As mentioned in [9], the reason for this limitation of the method is inherent in the approach. Since the first step is to identify (internal) equivalences and then proceed with an equivalence check treating these internal equivalences as *independent additional* inputs, the method may fail to recognize that certain circuits are identical. This failure to recognize equivalences is called *false negative*, in literature.

2. Partitioning the problem into a set of smaller and simpler equivalence checking problems

The decomposition based approach just described can be applied iteratively [9]. Let $\mathcal{N}$ be a miter with primary inputs $x_1, \ldots, x_n$. Now assume that there is a pair (s, t) of functionally equivalent signal lines. Such a pair of functionally equivalent signal lines is referred to as *cutpoint* [77]. The idea of the approach is to cut the circuit at the cutpoint as illustrated in Figure 6.2a and Figure 6.2b. That is, the circuit is cut at this position and the cutpoint is taken as additional primary input.

It is easy to see that the signal lines s and t of the circuit shown in Figure 6.2a are functionally equivalent. Both signal lines realize the Boolean function described by $x_1 \cdot x_2$. Thus, they can be merged. After the insertion of cutpoint $y_1 = (s, t)$ the new circuit $\mathcal{N}_{y_1}$ which is shown in Figure 6.2b realizes a Boolean function $\phi(\mathcal{N}_{y_1}) \in \mathcal{B}_{n+1}$ which depends on $x_1, \ldots, x_n$, and y_1. Variable y_1 is associated with the Boolean function $f_s \in \mathcal{B}_n$, i.e., the Boolean function which defines the behavior of signal line s subject to the primary inputs $x_1, \ldots, x_n$. In this context, we denote this Boolean function by χ_{y_1}.

Now, look once again at the circuit in Figure 6.2b. Now, the behavior of signal line u is defined by

$$f_u = {y_1}' + x_3$$

and the behavior of signal line v is defined by

$$\begin{aligned} f_v &= ((y_1 \cdot x_3)' \cdot y_1)' \\ &= (y_1 \cdot x_3) + {y_1}' \\ &= x_3 + {y_1}'. \end{aligned}$$

Thus, signal lines u and v are functionally equivalent and can be taken as further cutpoint y_2. Figure 6.2c shows the resulting circuit $\mathcal{N}_{\{y_1,y_2\}}$.

If k cutpoints $y_1, \ldots, y_k$ (we assume that the cutpoints are topologically ordered from the primary inputs to the primary output of the miter) are inserted, circuit $\mathcal{N}_{\{y_1,\ldots,y_k\}}$ realizes a Boolean function $\phi(\mathcal{N}_{\{y_1,\ldots,y_k\}}) \in \mathcal{B}_{n+k}$ over $x_1, \ldots, x_n, y_1, \ldots, y_k$. Each variable y_j is associated with its function $\chi_{y_j} \in \mathcal{B}_{n+j-1}$ over the variables $x_1, \ldots, x_n, y_1, \ldots, y_{j-1}$.

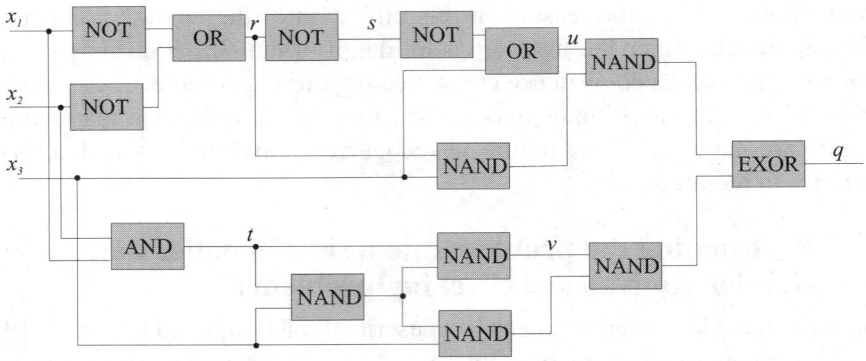

Figure 6.2a. Initial Boolean Network

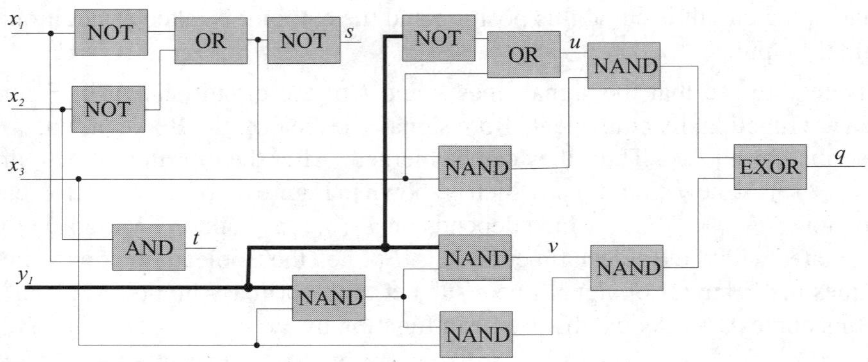

Figure 6.2b. Cutting the miter at cutpoint (s, t). After insertion of the cutpoint, the miter depends on x_1, x_2, x_3, and y_1.

If the primary output of $\mathcal{N}_{\{y_1,\ldots,y_k\}}$ is not satisfiable, then the circuits under comparison are functionally equivalent. If $\mathcal{N}_{\{y_1,\ldots,y_k\}}$ is satisfiable, we have to check whether there is a false negative.

Let us come back to the example above. Circuit $\mathcal{N}_{\{y_1,y_2\}}$ in Figure 6.2c is described by

$$
\begin{aligned}
\chi_{y_1} &= x_1 \cdot x_2, \\
\chi_{y_2} &= {y_1}' + x_3,
\end{aligned}
$$

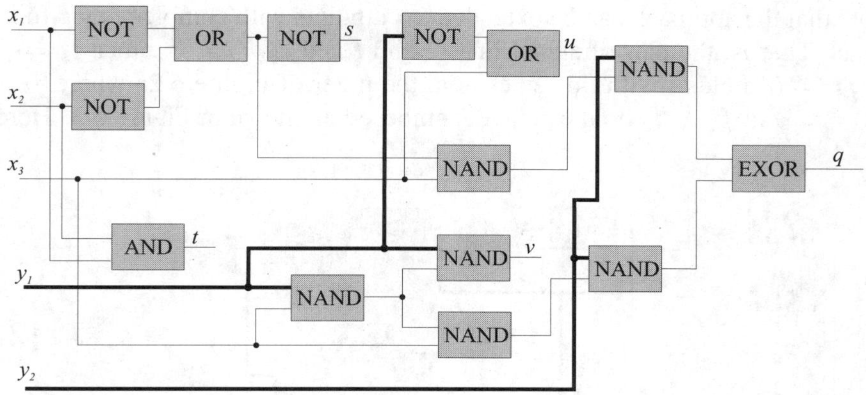

Figure 6.2c. Inserting cutpoints (s, t) and (u, v) into miter $\mathcal{N}$ of Figure 6.2a

and

$$
\begin{aligned}
f_q &= (y_2 \cdot (x_3 \cdot (x_1{}' + x_2{}'))')' \oplus (y_2 \cdot (x_3 \cdot (x_3 \cdot y_1)')')' \\
&= (y_2{}' + (x_3 \cdot (x_1{}' + x_2{}'))) \oplus (y_2{}' + (x_3 \cdot (x_3 \cdot y_1)')) \\
&= (y_2{}' + (x_3 \cdot (x_1{}' + x_2{}'))) \oplus (y_2{}' + (x_3 \cdot (x_3{}' + y_1{}'))) \\
&= (y_2{}' + (x_1{}' \cdot x_3) + (x_2{}' \cdot x_3)) \oplus (y_2{}' + (x_3 \cdot y_1{}')).
\end{aligned}
$$

Unfortunately, the primary output q of $\mathcal{N}_{\{y_1,y_2\}}$ is satisfiable since f_q which depends on x_1, x_2, x_3, y_1, and y_2 is obviously different from the constant Boolean function **0**. Just assign the values 1, 1, 1, 0, and 1 to the variables x_1, x_2, x_3, y_1, and y_2, respectively. Thus, we have to check whether there is a false negative. And indeed, there is a false negative since resubstitution of variable y_1 by χ_{y_1} results in

$$
\begin{aligned}
&(f_q)_{y_1 = x_1 \cdot x_2} \\
&= ((y_2{}' + (x_1{}' \cdot x_3) + (x_2{}' \cdot x_3)) \oplus (y_2{}' + (x_3 \cdot y_1{}')))_{y_1 = x_1 \cdot x_2} \\
&= (y_2{}' + (x_1{}' \cdot x_3) + (x_2{}' \cdot x_3)) \oplus (y_2{}' + (x_3 \cdot (x_1 \cdot x_2)')) \\
&= (y_2{}' + (x_1{}' \cdot x_3) + (x_2{}' \cdot x_3)) \oplus (y_2{}' + (x_3 \cdot (x_1{}' + x_2{}'))) \\
&= (y_2{}' + (x_1{}' \cdot x_3) + (x_2{}' \cdot x_3)) \oplus (y_2{}' + (x_1{}' \cdot x_3) + (x_2{}' \cdot x_3)) \\
&= \mathbf{0}.
\end{aligned}
$$

This proves that the two circuits under consideration are functionally equivalent. The false negative has been generated by the fact that in the initial miter shown in Figure 6.2a the signal lines s and t are 1 whenever $x_1 = x_2 = 1$ holds, i.e., the assignment 1, 1, and 0 to the primary inputs x_1, x_2, and y_1, respectively, is not a consistent assignment.

Note that the approach can be extended to cutpoints with complemented functions. That is, the pair of signal lines s and t is a *cutpoint* if either $f_s = f_t$ or $f_s = f_t{}'$ holds. With this extension, the miter of Figure 6.2a where $f_t = f_s = f_r{}'$ and $f_u = f_v$ hold can be decomposed as shown in Figure 6.3. Here,

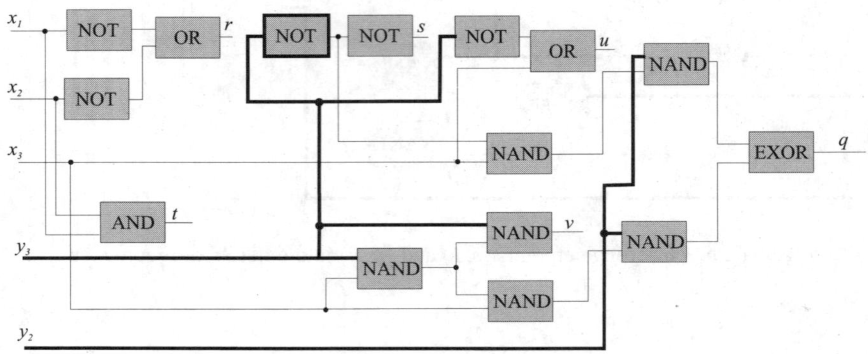

Figure 6.3. Using cutpoints with complemented functions

equivalence of the two circuits being compared can directly be shown because of

$$\begin{aligned} f_q &= (y_2 \cdot (x_3 \cdot y_3{}')')' \oplus (y_2 \cdot (x_3 \cdot (x_3 \cdot y_3)')')' \\ &= (y_2{}' + (x_3 \cdot y_3{}')) \oplus (y_2{}' + (x_3 \cdot (x_3 \cdot y_3)')) \\ &= (y_2{}' + (x_3 \cdot y_3{}')) \oplus (y_2{}' + (x_3 \cdot (x_3{}' + y_3{}'))) \\ &= (y_2{}' + (x_3 \cdot y_3{}')) \oplus (y_2{}' + (x_3 \cdot y_3{}')) \\ &= \mathbf{0}. \end{aligned}$$

Basically, the overall procedure for *cutpoint based equivalence checking* consists of two phases, namely

- detection and selection of cutpoints, and
- false negative resolution in case of mismatch.

Resubstitution of cutpoints y_i by their functions χ_{y_i} as shown in the example on page 135 is used to solve the false negative resolution problem, in practice [9, 77]. However, the method is very sensitive to the order in which the cutpoints are resubstituted. As mentioned in [77], a bad order might cause the elimination of all cutpoints including the ones which do not cause false negatives. A BDD based approach to false negative resolution which uses resubstitution can be found in [77]. The false negative resolution problem can also be attacked by the approach of Gupta and Ashar [55] which is presented in Chapter 5 on page 116.

Since the first phase is applied within the context of another approach for equivalence checking as well (see Section 3), we separately handle it in Section 4.

3. Using internal equivalences to simplify large miters

Cutpoint based equivalence checkers have a big drawback, the false negatives. Whenever cutpoint based equivalence checkers fail to recognize that the circuits under consideration are functionally identical, one has to check whether there is a false negative. As already mentioned the method of resubstitution of cutpoints is very sensitive to the order in which the cutpoints are resubstituted. A bad order might cause the elimination of all cutpoints including the ones which do not cause false negatives. Furthermore, proving a faulty Boolean network to be faulty results in many resubstitutions in general, as no false negative can be found.

False negatives are due to the fact that taking cutpoints as additional primary inputs often oversimplifies the problem instance. A more natural approach is to use cutpoints, i.e., functionally equivalent internal signal lines, to simplify the miter under consideration *without loosing information* as proposed by Brand [16].

Let us assume that the signal lines s and t of the miter in Figure 6.4 have been proven to be functionally equivalent. Then, the miter can be simplified by transforming it as shown in the same figure: The gate inputs which are driven by signal line t are connected to signal line s; gates which now are superfluous are deleted. Thus, after the transformation the circuits being compared share the logic which drives signal line s and the reconvergences of the miter, i.e., the distance between the position where the signals diverge and the position where the signals converge, have usually been reduced. As explained in [16], the approach can be extended to signal lines s and t with $f_s = f_t{}'$. In this case, the complemented signal line s can replace signal line t.

Let us illustrate the iterative process proposed by [16] by an example (see Figures 6.5a-6.5e). The circuit under consideration has already been used in the section above. We consider the signal lines in a topological order. The first pair of "equivalent" signal lines are signal lines j and p. Complementing signal line j results in a signal which is functionally equivalent to signal p. Thus, we can "replace signal p" by the complemented signal line j. The resulting miter which is shown in Figure 6.5b is smaller than the initial miter. Due to this simplification of the miter, it may be easier to find further (internal) signal equivalences in the miter.

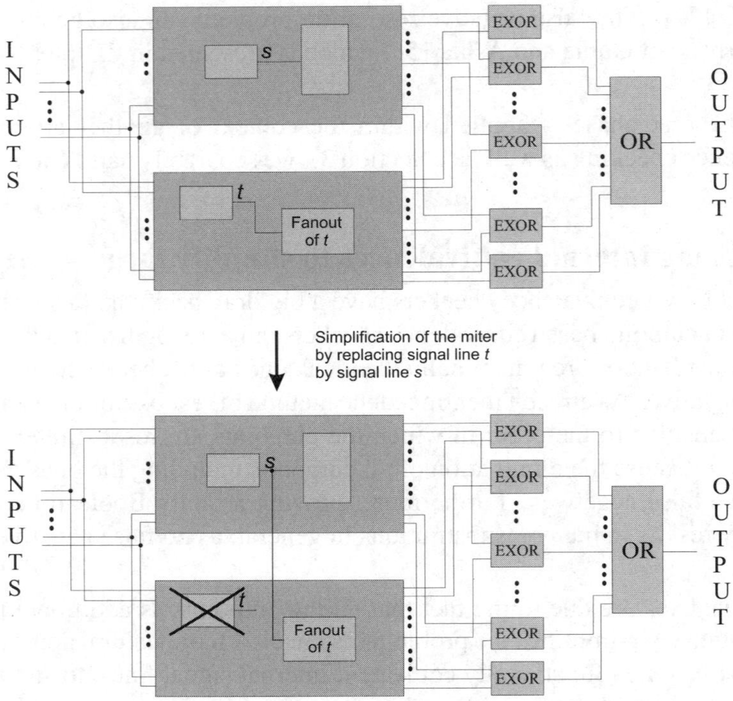

Figure 6.4. Using internal equivalences to simplify large miters.

For signal lines g and l, the equalities

$$\begin{aligned} f_l &= (j \cdot k)' \\ &= (j \cdot (j \cdot x_3)')' \\ &= (j \cdot (j' + x_3'))' \\ &= (j \cdot x_3')' \\ &= j' + x_3 \\ &= p + x_3 \\ &= f_g \end{aligned}$$

hold. Thus, g and l are functionally equivalent signal lines and signal line g can be replaced by signal line l (see Figure 6.5c). Analogously, it can be shown that signal lines h and m are functionally equivalent and signal line h can be replaced by signal line m. The resulting miter is shown in Figure 6.5d. Obviously, the signals i and n which drive the inputs of the EXOR-gate are

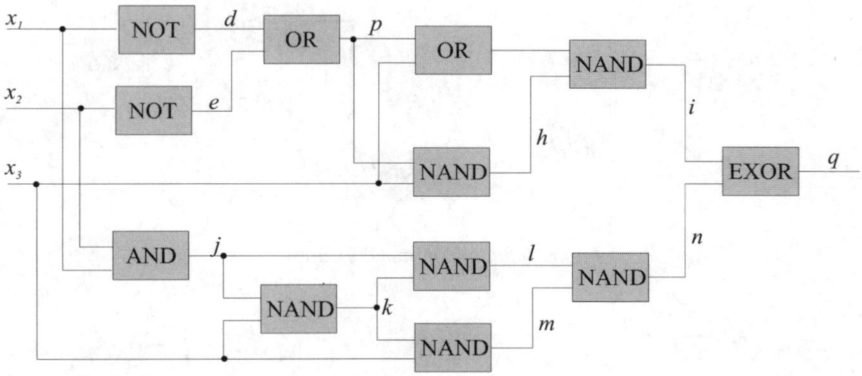

Figure 6.5a. Example illustrating the approach of Brand [16]: The initial circuit (which has been taken from [77, 78]).

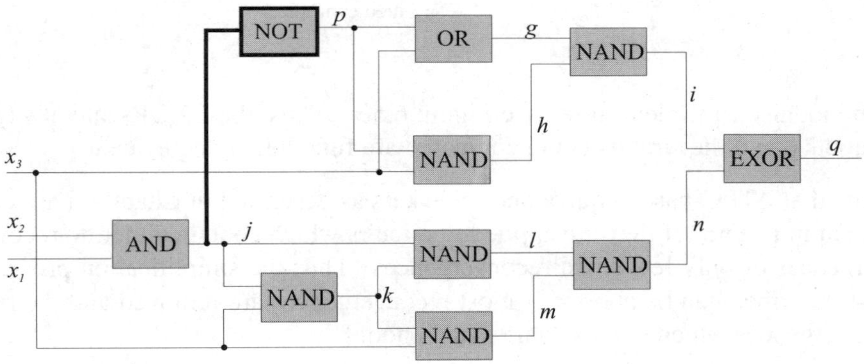

Figure 6.5b. Replacement of signal line p by the complemented signal line j.

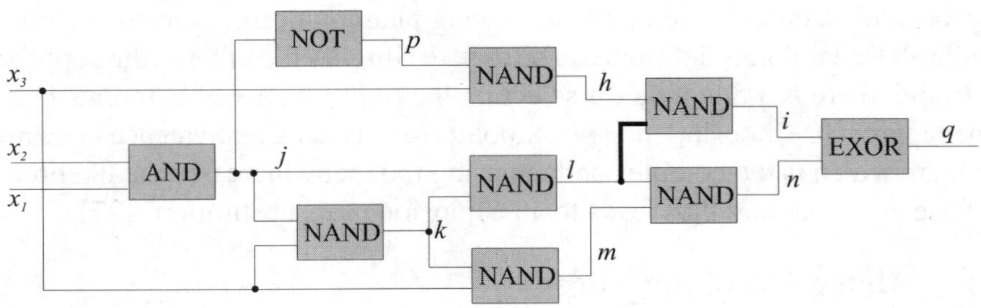

Figure 6.5c. Replacement of signal line g by the complemented signal line l.

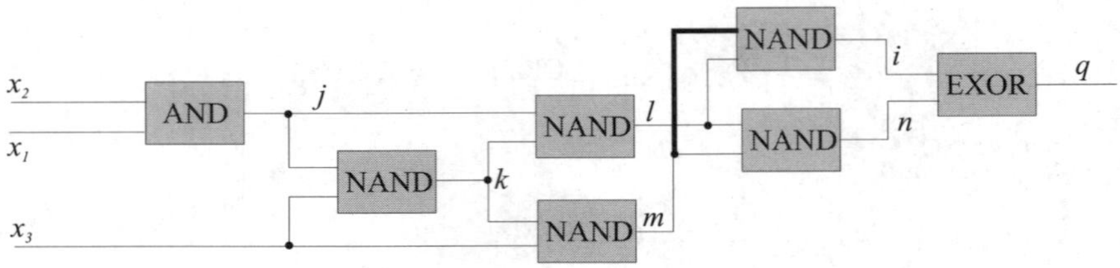

Figure 6.5d. Continued example.

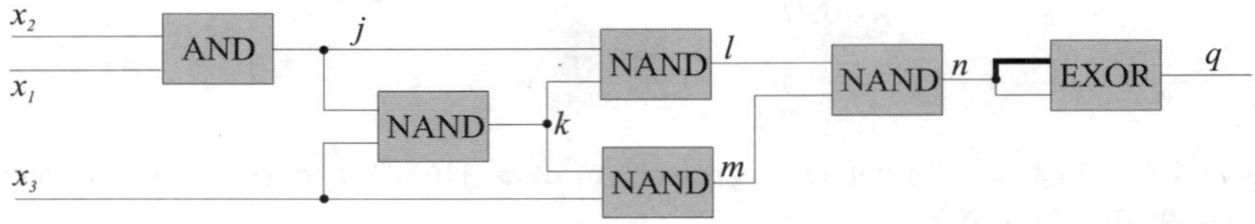

Figure 6.5e. Continued example.

functionally equivalent, now (see Figure 6.5e). Thus, the EXOR-gate always outputs 0 and the circuits being compared are functionally equivalent.

Note that ATPG-based equivalence checkers as described in Chapter 5 are efficient in general, if they are applied to circuits which contain either no reconvergences or only few small reconvergences. Thus the simplification process just described can be aborted if most reconvergences are removed and the reconvergences which still exist are small enough.

4. Finding internal equivalences

Both the approach of Brand which has been presented in Section 3 and the approach of Section 2 need a preprocessing phase which computes cutpoints. While all the cutpoints detected can be used to simplify the miter in the approach of Brand, there is a dilemma for selecting the right number of cutpoints in the other approach. Choosing too few cutpoints results in an equivalence checking problem which is too complex and choosing too many increases the likelihood of false negatives and thus leads to an explosion of resubstitutions [77].

4.1 Detection of cutpoints

Let us begin with algorithms to detect cutpoints.

4.1.1 Filtering phase

As proposed in [9], in a first step candidates for cutpoints are computed. Usually, this is done by random simulation, i.e., the miter is simulated for a small number of random assignments to the primary inputs. These simulation runs produce a vector ω_s of values at each signal line s of the miter. This vector can be used as *signature* of signal line s. Whenever, neither $\omega_s = \omega_t$ nor $\omega_s = \omega_t{}'$ hold, the pair (s,t) cannot be used as cutpoint. Thus, this step gives us a list of pairs (s,t) of signal lines which are potential candidates for internal functional equivalences. The list is ordered in a breadth-first manner such that "earlier pairs" come first. Now, the pairs contained in this list are taken one after the other and the corresponding signal lines s and t are checked for functional equivalence.

4.1.2 Naive approach

A first idea to implement this check is either

- to compute the ROBDDs of f_s and f_t by symbolic simulation (see Chapter 4 Section 1.1 on page 72)

or

- to insert an EXOR-gate whose inputs are connected to s and t, respectively, and to ask whether the output of the EXOR-gate, *which is treated as additional primary output of the circuit*, is not testable for stuck-at-1 or stuck-at-0 (see Chapter 5 Section 2.2 on page 122). If it is not testable for stuck-at-0 then $f_s = f_t$. If it is not testable for stuck-at-1 then $f_s = f_t{}'$.

Figure 6.6 illustrates the latter approach.

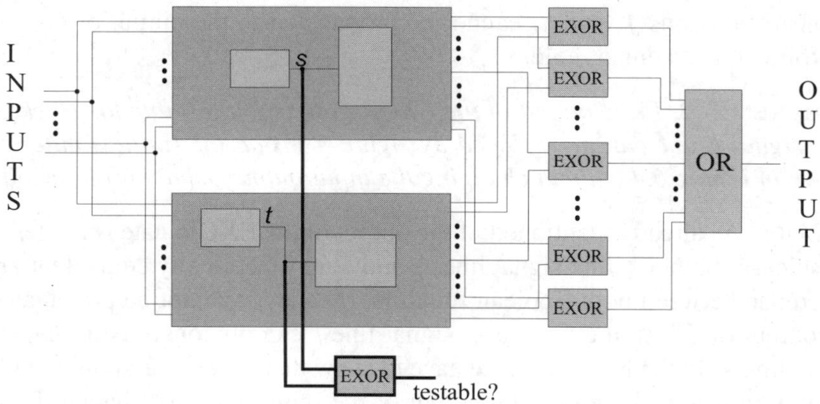

Figure 6.6. First idea for finding internal equivalences.

The problem with both ideas is that they fail if the internal nodes of the circuit implementation are optimized with respect to local don't cares so that the miter does not contain many functionally equivalent signal lines (cf. [16]).

4.1.3 Approach of Brand

A more sophisticated approach has been proposed in [16]. Here, an EXOR-gate is inserted between the signal line t and all its immediate fanouts. The other input of the EXOR-gate is connected to signal line s. Let us denote this transformed circuit by C. Figure 6.7 shows the circuit C just described.

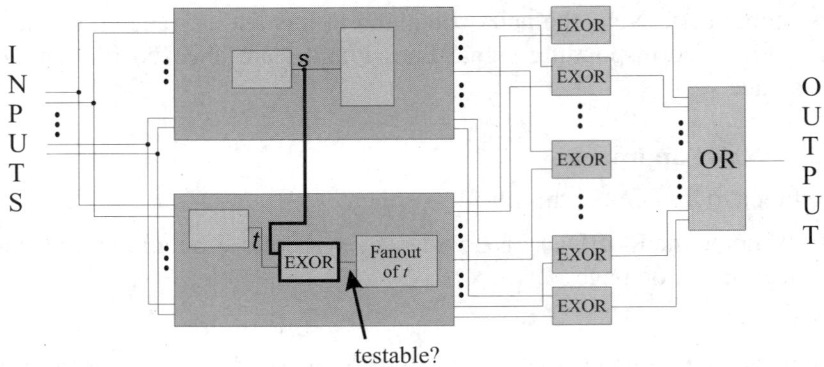

Figure 6.7. Circuit C obtained by inserting an EXOR-gate between the signal line t and its immediate fanouts and connecting the other input of the EXOR-gate to signal line s.

Now, if the output of the EXOR-gate is not testable for stuck-at-0, then either signal lines s and t are functionally equivalent or the difference between both Boolean functions f_s and f_t cannot be propagated to the output of C. Thus, the following statement holds.

THEOREM 6.2 *If the output of the EXOR-gate is not testable for stuck-at-0, then signal line t can be replaced by signal s inside the (initial) miter $\mathcal{N}$ as shown in Figure 6.4 without changing the input/output behavior of the miter.*

PROOF: As already mentioned, if the output of the EXOR-gate is not testable for stuck-at-0, then either signal lines s and t are functionally equivalent or the difference between both Boolean functions f_s and f_t cannot be propagated to the output of C. In the first case, signal line t can obviously be replaced by signal line s. In the latter case, we have to show that for each assignment to the primary inputs of the miter, the output of the miter does not change if signal line t is replaced by s. To prove it, let vector α be an assignment to the primary inputs of the miter such that $f_s(\alpha) \neq f_t(\alpha)$.

- At first, let us assume that $f_s(\alpha) = 0$ and $f_t(\alpha) = 1$ hold. Then, the immediate fanouts of t are assigned to value 1 in the initial miter $\mathcal{N}$ and to value 0 in the simplified miter. However, this is the situation which occurred when we searched for a test for stuck-at-0 at the output of the EXOR-gate inserted in circuit C: The output of the EXOR-gate has to be assigned value 1 in the faultless circuit C and value 0 in the faulty circuit C. Since such a test does not exist, it cannot be detected at the output of the miter under input assignment α whether value 0 or value 1 is assigned to the immediate fanouts of t.

- Now, let us assume that $f_s(\alpha) = 1$ and $f_t(\alpha) = 0$ hold, i.e., the immediate fanouts of t are assigned to value 0 in the initial miter $\mathcal{N}$ and to value 1 in the simplified miter. Obviously, this situation is the same one as above.

■

Similarly, if the output of the EXOR-gate inserted is not testable for stuck-at-1, then the complemented signal line s can replace signal line t.

4.2 Selection of cutpoints

As already mentioned above, we have to select a proper subset of the detected cutpoints when applying the decomposition based approach of Section 2, as there is a dilemma for selecting the right number of cutpoints in the cutpoint based approach. Choosing too few cutpoints results in an equivalence checking problem which is too complex and choosing too many increases the likelihood of false negatives and thus to an explosion of resubstitutions [77].

Several heuristics have been proposed to attack this problem. We will mention some of them very briefly, in the following:

- *Select a subset of mutually independent cutpoints* [98], whereby a signal line s is said to depend on signal line t if signal line t is contained in the transitive fanin of signal line s. This approach is justified by the observation that false negatives occur if mutually dependent cutpoints are selected.

- *Select cutpoints with high fanout* [140]. Selecting a cutpoint with a high fanout instead of selecting a cutpoint with low fanout induces that, after cutting the miter, more signal lines depend on the same (new) variable which makes false negatives more improbable.

- *Select cutpoints only if the amount of logic between two cutpoints is not too small* [140]. This rule is due to the fact that fine grained decompositions lead to an unnecessary large number of variables, which increases the running time of the underlying algorithms, and run an increased risk of false negatives.

Chapter 7

CHECKING EQUIVALENCE FOR PARTIAL IMPLEMENTATIONS

Guest Authors[1]
Christoph Scholl, University Heidelberg, Germany
Bernd Becker, University Freiburg, Germany

In this chapter we address the problem of *Black Box Equivalence Checking*, which occurs when the specification is known, but only parts of the implementation are finished or known. For an example, see Figure 7.1a and Figure 7.1b. The latter one shows a partial implementation of the specification given in Figure 7.1a. It contains two black boxes, BB_1 and BB_2. Clearly, after a suitable implementation of the two black boxes the final implementation fulfills its specification.

Black Box Equivalence Checking enables the use of verification techniques in early stages of the design. Design errors can be already detected when only a partial implementation is at hand – for example, due to a distribution of the implementation task to several groups of designers. Parts of the implementation, which are not yet finished, are combined into *black boxes*. If the implementation differs from the specification for all possible substitutions of the black boxes, a design error is found in the current partial implementation. To detect an error in the current partial implementation it is necessary to find an assignment to the primary inputs which produces erroneous values at the outputs independently from the final implementation of the black boxes.

Another application of *Black Box Equivalence Checking* is the abstraction from "difficult parts" of an implementation, which would cause a large peak size in memory consumption during the construction of a canonical form for the implementation. These "difficult parts" of the design can be put into a black box and Black Box Equivalence Checking is performed. An exact statement about the correctness of the full implementation is not possible, but it is still

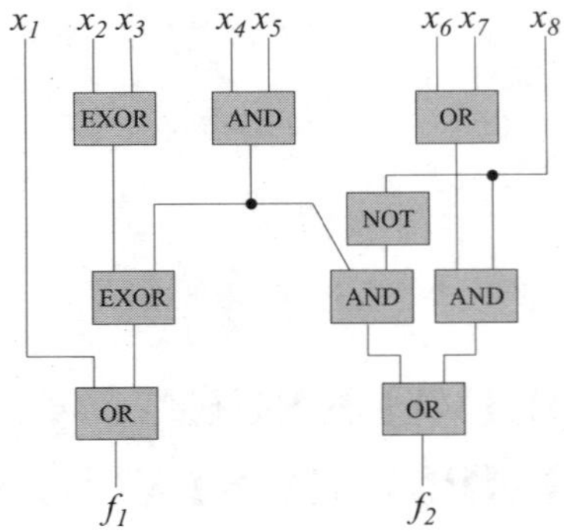

Figure 7.1a. Specification and partial implementation: Specifying circuit

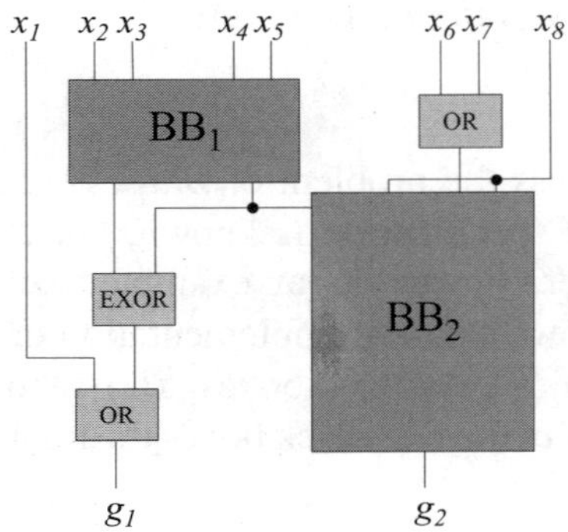

Figure 7.1b. Partial implementation with two black boxes

possible to find errors in the partial implementation given to the Black Box Equivalence Checker.

Black Box Equivalence Checking can also be used to verify assumptions concerning the location of errors in implementations, which do not fulfill their specifications. If there is some assumption on the location of errors (produced by an automatic error diagnosis tool or found by hand), then these regions of the design are cut off and put into black boxes. If Black Box Equivalence Checking gives the information that no error can be found in the design containing black boxes, we can conclude that the assumptions on the error location were correct, otherwise we know that there must be errors also in other regions of the design.

In this chapter, we present algorithms for equivalence checking of partial implementations under the assumption that a combinational circuit is given as specification and all implementations and black boxes are of combinational nature. We present a thorough analysis of the problem leading to several algorithms to attack the Black Box Equivalence Checking problem. These algorithms need different amounts of resources in terms of space and time and differ from their accurateness. They range from a simple algorithm using symbolic simulation for an approximation of the solution to an exact solution of the problem. Approximate solutions are not able to find all errors in the partial implementation, but they are correct in the sense that they do not report an error if there is still a possibility to implement the black boxes leading to a correct overall implementation. However, if we solve the Black Box Equivalence Checking problem approximatively, the information, that no error can be found, can be due to the approximative character of the approach and does not necessarily imply that there is an implementation of the black boxes leading to a correct overall implementation. That is, when Black Box Equivalence Checking is used to verify assumptions on the location of design errors, it cannot be guaranteed that the information, that no error can be found, implies that the error location is confined to the black boxes.

We start with a simple symbolic simulation with respect to the $0, 1, X$ logic. Then we successively increase the exactness (and the complexity) of the algorithm leading to a *local check*, an *output exact check* and an *input exact check*. In particular, an exact criterion to decide for a given partial implementation with *one* black box and a specification whether the partial implementation is correct or not is given. Unlike the approaches presented in [54] and [67], it is guaranteed that there is really an extension of the partial implementation to a correct complete implementation, if the criterion which is proven in Section 2.3.1 on page 161 reports no error (and of course, vice versa, there is no extension of the partial implementation to a complete implementation, if it does report an error). Several experiments have been performed which show that improving the accuracy of the algorithms indeed leads to a significant improvement of the error detection capabilities (paid with an increase of computational resources). They can be found in [130].

In the following, we assume that the circuit under consideration has n input signal lines, $x_1, \ldots, x_n$. As a running example for the demonstration of our algorithms we will often use the specification given in Figure 7.1a.

1. Symbolic Z–simulation

A first algorithm for checking partial implementations is based on the usual $\{0, 1, X\}$-simulation, which is well-known in the area of testing [1] (cf. Chapter 5 Section 2.2.1).

To simulate a partial implementation with n primary inputs for an input vector $(\alpha_1, \ldots, \alpha_n) \in \{0, 1\}^n$ the unknown value X is assigned to all outputs of the black boxes. If all the inputs of a gate are assigned a value in $\{0, 1\}$, then the output of the gate is computed according to the gate function as usual. If some inputs of a gate are set to X, the output is equal to X if and only if there are two different replacements of the X values at the inputs by 0's and 1's, which lead to different outputs of the gate. For a two-input AND-gate, a two-input OR-gate, a two-input $EXOR$-gate, and an inverter, this leads to the rules

x_1	x_2	$\cdot$	$+$	$\oplus$	$x_1{}'$
0	0	0	0	0	1
0	1	0	1	1	1
0	X	0	X	X	1
1	1	1	1	0	0
1	X	X	1	X	0
X	X	X	X	X	X.

Figure 7.2 shows the evaluation of the partial implementation of Figure 7.1b with respect to the input vector $(1, 0, 0, 0, 0, 0, 0, 0)$. Note that the first output of the partial implementation is 1 independently of the functionality of the black boxes.

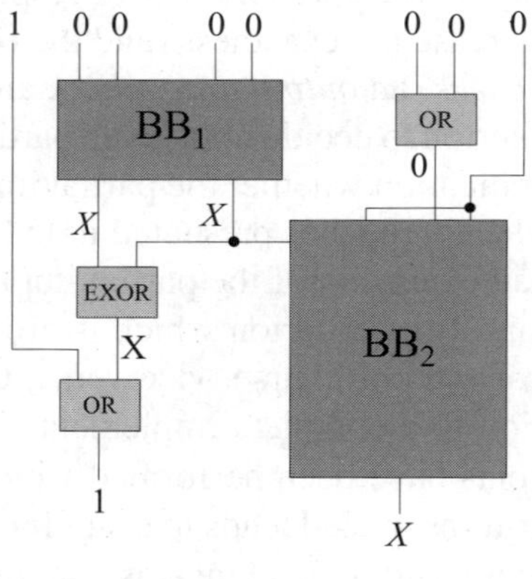

Figure 7.2. Evaluation with respect to input vector $(1, 0, 0, 0, 0, 0, 0, 0)$

We can take advantage of such a $\{0, 1, X\}$-simulation to detect errors in partial implementations. If the evaluation of the partial implementation results in a value $\beta \in \{0, 1\}$ for some output, this means that the output value is β independently of the functionality of the black boxes. If on the other hand the specification produces β' for the same input vector, then we have found an error in the partial implementation. Figure 7.3 shows such a situation. When this partial implementation is compared to the specification of Figure 7.1a by applying input vector $(1, 0, 0, 0, 0, 0, 0, 0)$, we see that the first output of the (partial) implementation is 0 whereas it is 1 for the specification. Generalizing the usual

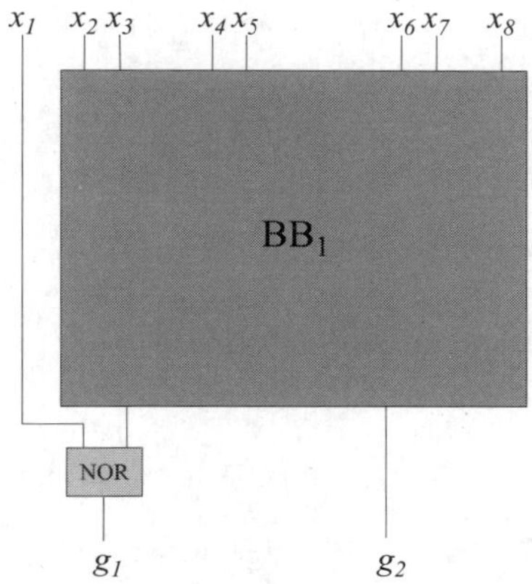

Figure 7.3. Partial implementation with a detectable error

notion of a *distinguishing vector* for designs without black boxes to designs containing black boxes we can say that $(1, 0, 0, 0, 0, 0, 0, 0)$ is a *distinguishing vector* for the specification in Figure 7.1a and the partial implementation in Figure 7.3. Of course, only vectors which produce 0 or 1 (i.e., which do not produce X) at outputs of the partial implementation can play a role as distinguishing vectors.

Since it is not possible to simulate specification and implementation for all 2^n input vectors one after the other to find distinguishing vectors, a more sophisticated method has to be applied. For this, consider an output i of the partial implementation and the Boolean function $g_i \in \mathcal{B}_{n+1}$ with $n + 1$ Boolean variables $x_1, \ldots, x_n$, and Z which is defined by Property I.

> **Property I** For each assignment $\alpha = (\alpha_1, \ldots, \alpha_n) \in \{0, 1\}^n$ of the primary inputs of the partial implementation, the iterated cofactor
>
> $$g_{i[x_1,\ldots,x_n]=(\alpha_1,\ldots,\alpha_n)}$$
>
> of g_i with respect to $x_1 = \alpha_1, \ldots, x_n = \alpha_n$ is the constant Boolean function **0** or **1** if the $\{0, 1, X\}$-simulation assigns value 0 or value 1, respectively, to output i. It is the Boolean function $\boldsymbol{Z} \in \mathcal{B}_{n+1}$ defined by
>
> $$\boldsymbol{Z}(x_1, \ldots, x_n, Z) = Z$$
>
> if the $\{0, 1, X\}$-simulation assigns value X to output i.

Thus, function g_i "characterizes" the outputs of the $\{0, 1, X\}$-simulation runs applied to the partial implementation. Figure 7.4 and Figure 7.5 illustrate the

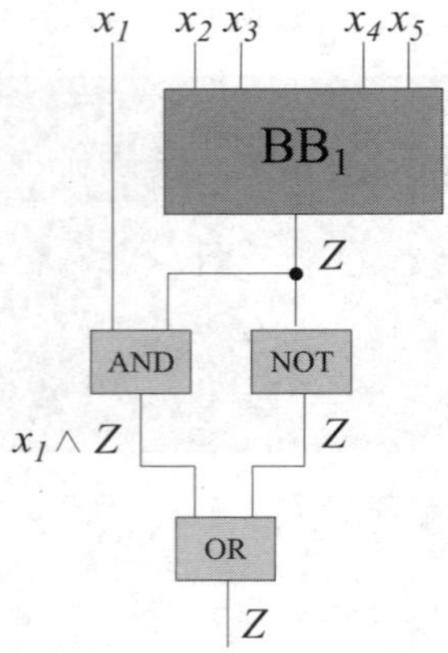

Figure 7.4. Illustration of symbolic Z-simulation

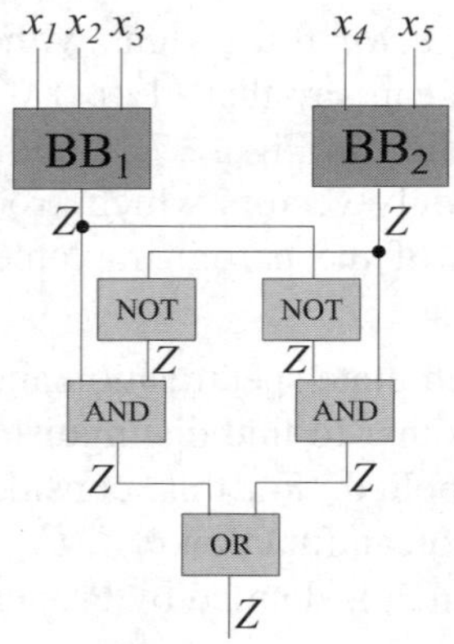

Figure 7.5. Symbolic Z-simulation of an EXOR-gate whose input signal lines are driven by black boxes

definition. We have tagged each signal line j of the circuit with the corresponding "g_j-function".

The computation of the ROBDDs of the g_i's is quite easy. It is performed by so called *symbolic Z-simulation*. Since all types of gates can be expressed using two-input AND-gates, two-input OR-gates, and inverters, let us assume without loss of generality that all our gates have types $\cdot$, $+$, or NOT to explain the approach, in detail. Firstly, the primary input signal lines of the circuit are associated with unique ROBDD variables, $x_1, \ldots, x_n$. All output signals of black boxes are associated with the new variable Z. Now, ROBDDs for the functions computed by the gates of the circuit are built in topological order treating the black box outputs (associated with variable Z) as additional primary inputs of the circuit. The gates of the circuit can be processed in a manner similar to a conventional symbolic simulation which has been presented in Chapter 3 Section 3.2.1:

- When an AND- or an OR-gate is processed, the ROBDDs P and Q of the two predecessor functions are combined by computing $ITE(P, Q, \mathbf{0})$ or $ITE(P, P, Q)$, respectively.

- For an inverter, the NOT operation on the ROBDD of the predecessor function has to be performed, followed by a substitution operation which composes Z' for Z.

The fact that the final ROBDD obtained by this modified symbolic simulation represents Boolean function g_i follows from the observation that the initial ROBDDs for the primary inputs and for the black box outputs fulfill Property I and from the following three equalities:

1. Let p and q be the two functions computed for the predecessor gates of an AND-gate. By induction, we can assume that both fulfill Property I. Thus, for each assignment $(\alpha_1, \ldots, \alpha_n)$ to the Boolean variables $x_1, \ldots, x_n$, the equation

$$
\begin{aligned}
& (p \cdot q)_{[x_1,\ldots,x_n]=(\alpha_1,\ldots,\alpha_n)} \\
= \; & p_{[x_1,\ldots,x_n]=(\alpha_1,\ldots,\alpha_n)} \cdot q_{[x_1,\ldots,x_n]=(\alpha_1,\ldots,\alpha_n)} \\
= \; & \begin{cases}
0, \text{ if } p_{[x_1,\ldots,x_n]=(\alpha_1,\ldots,\alpha_n)} = 0 \text{ or } q_{[x_1,\ldots,x_n]=(\alpha_1,\ldots,\alpha_n)} = 0 \\
1, \text{ if } p_{[x_1,\ldots,x_n]=(\alpha_1,\ldots,\alpha_n)} = 1 \text{ and } q_{[x_1,\ldots,x_n]=(\alpha_1,\ldots,\alpha_n)} = 1 \\
Z, \text{ if } \{p_{[x_1,\ldots,x_n]=(\alpha_1,\ldots,\alpha_n)}, q_{[x_1,\ldots,x_n]=(\alpha_1,\ldots,\alpha_n)}\} = \{Z\} \\
\quad \text{or } \{p_{[x_1,\ldots,x_n]=(\alpha_1,\ldots,\alpha_n)}, q_{[x_1,\ldots,x_n]=(\alpha_1,\ldots,\alpha_n)}\} = \{1, Z\}
\end{cases}
\end{aligned}
$$

holds. Thus, the Boolean function $(p \cdot q)_{[x_1,\ldots,x_n]=(\alpha_1,\ldots,\alpha_n)}$ equals function $\boldsymbol{Z}$ if and only if the corresponding $\{0, 1, X\}$-simulation assigns either value X to both input signal lines of the AND-gate or value X to one of the input signal lines and value 1 to the other one. That is, function $p \cdot q$ fulfills Property I, too.

2. Let p and q be the two functions computed for the predecessor gates of an OR-gate. By induction, we can assume that both fulfill Property I. Thus, for each assignment $(\alpha_1, \ldots, \alpha_n)$ to the Boolean variables $x_1, \ldots, x_n$, the equation

$$
\begin{aligned}
&(p+q)_{[x_1,\ldots,x_n]=(\alpha_1,\ldots,\alpha_n)} \\
&= p_{[x_1,\ldots,x_n]=(\alpha_1,\ldots,\alpha_n)} + q_{[x_1,\ldots,x_n]=(\alpha_1,\ldots,\alpha_n)} \\
&= \begin{cases} 0, \text{ if } p_{[x_1,\ldots,x_n]=(\alpha_1,\ldots,\alpha_n)} = 0 \text{ and } q_{[x_1,\ldots,x_n]=(\alpha_1,\ldots,\alpha_n)} = 0 \\ 1, \text{ if } p_{[x_1,\ldots,x_n]=(\alpha_1,\ldots,\alpha_n)} = 1 \text{ or } q_{[x_1,\ldots,x_n]=(\alpha_1,\ldots,\alpha_n)} = 1 \\ Z, \text{ if } \{p_{[x_1,\ldots,x_n]=(\alpha_1,\ldots,\alpha_n)}, q_{[x_1,\ldots,x_n]=(\alpha_1,\ldots,\alpha_n)}\} = \{Z\} \\ \quad \text{or } \{p_{[x_1,\ldots,x_n]=(\alpha_1,\ldots,\alpha_n)}, q_{[x_1,\ldots,x_n]=(\alpha_1,\ldots,\alpha_n)}\} = \{0, Z\} \end{cases}
\end{aligned}
$$

holds. Thus, function $p + q$ fulfills Property I, too.

3. Let p be the function computed for the predecessor gate of an inverter. By induction, we can assume that it fulfills Property I. Thus, for each assignment $(\alpha_1, \ldots, \alpha_n)$ to the Boolean variables $x_1, \ldots, x_n$, the equation

$$
\begin{aligned}
&((p')_{Z=Z'})_{[x_1,\ldots,x_n]=(\alpha_1,\ldots,\alpha_n)} \\
&= \left((p')_{[x_1,\ldots,x_n]=(\alpha_1,\ldots,\alpha_n)}\right)_{Z=Z'} \\
&= \left((p_{[x_1,\ldots,x_n]=(\alpha_1,\ldots,\alpha_n)})'\right)_{Z=Z'} \\
&= \begin{cases} 0 \quad, \text{ if } p_{[x_1,\ldots,x_n]=(\alpha_1,\ldots,\alpha_n)} = 1 \\ 1 \quad, \text{ if } p_{[x_1,\ldots,x_n]=(\alpha_1,\ldots,\alpha_n)} = 0 \\ (Z')', \text{ if } p_{[x_1,\ldots,x_n]=(\alpha_1,\ldots,\alpha_n)} = Z \end{cases} \\
&= \begin{cases} 0, \text{ if } p_{[x_1,\ldots,x_n]=(\alpha_1,\ldots,\alpha_n)} = 1 \\ 1, \text{ if } p_{[x_1,\ldots,x_n]=(\alpha_1,\ldots,\alpha_n)} = 0 \\ Z, \text{ if } p_{[x_1,\ldots,x_n]=(\alpha_1,\ldots,\alpha_n)} = Z \end{cases}
\end{aligned}
$$

holds.

After we have obtained functions g_i for all outputs of the partial implementation by symbolic Z-simulation and functions f_i for all outputs of the specification by a conventional symbolic simulation, the check whether there is a distinguishing vector between specification and implementation is based on the following two lemmas.

LEMMA 7.1 *There is no input vector* $(\alpha_1, \ldots, \alpha_n) \in \{0,1\}^n$ *with*

$$(g_i)_{[x_1,\ldots,x_n]=(\alpha_1,\ldots,\alpha_n)} = 1 \text{ and } f_i(\alpha_1, \ldots, \alpha_n) = 0$$

if and only if

$$((g_i)_{Z=0} \Rightarrow f_i) = 1.$$

PROOF: Let us assume without loss of generality that variable Z is the last variable in the variable order of the ROBDD (without complemented edges) of g_i. As each cofactor of g_i with respect to the primary input variables $x_1, \ldots, x_n$ is either **0**, **1**, or $\boldsymbol{Z}$, all the nodes of the ROBDD of g_i labelled with variable Z represent the Boolean function $\boldsymbol{Z}$. Thus, for all vectors $(\alpha_1, \ldots, \alpha_n) \in \{0,1\}^n$ the implication

$$(g_i)_{Z=0}(\alpha_1, \ldots, \alpha_n) = 1 \Rightarrow (g_i)_{[x_1,\ldots,x_n]=(\alpha_1,\ldots,\alpha_n)} = 1$$

holds.

Now assume that

$$((g_i)_{Z=0} \Rightarrow f_i) \neq 1$$

holds. That is, there is an assignment $(\alpha_1, \ldots, \alpha_n)$ to the primary input variables $(x_1, \ldots, x_n)$ such that

$$(g_i)_{Z=0}(\alpha_1, \ldots, \alpha_n) = 1 \text{ and } f_i(\alpha_1, \ldots, \alpha_n) = 0$$

which implies

$$(g_i)_{[x_1,\ldots,x_n]=(\alpha_1,\ldots,\alpha_n)} = 1 \text{ and } f_i(\alpha_1, \ldots, \alpha_n) = 0.$$

This finishes the proof of the if-part of the statement.

Now suppose that

$$((g_i)_{Z=0} \Rightarrow f_i) = 1$$

holds. Then we have to consider two cases for an arbitrary input vector $(\alpha_1, \ldots, \alpha_n) \in \{0,1\}^n$:

- If $(g_i)_{Z=0}(\alpha_1, \ldots, \alpha_n) = 0$, then $g_i(\alpha_1, \ldots, \alpha_n, 0) = 0$. Consequently, $(g_i)_{[x_i,\ldots,x_n]=(\alpha_1,\ldots,\alpha_n)} \neq 1$ holds.
- If $(g_i)_{Z=0}(\alpha_1, \ldots, \alpha_n) = 1$, we can conclude $f_i(\alpha_1, \ldots, \alpha_n) = 1$.

This completes the proof. ▪

LEMMA 7.2 *There is no input vector* $(\alpha_1, \ldots, \alpha_n) \in \{0,1\}^n$ *with*

$$(g_i)_{[x_1,\ldots,x_n]=(\alpha_1,\ldots,\alpha_n)} = 0 \textit{ and } f_i(\alpha_1, \ldots, \alpha_n) = 1$$

if and only if

$$((g_i')_{Z=1} \Rightarrow f_i') = 1.$$

PROOF: The statement can be proved analogously to Lemma 7.1. ▪

2. Symbolic Z_i-simulation

A disadvantage of symbolic Z-simulation lies in the fact that not all errors, which are present in a partial implementation, can be found by the procedure described above.

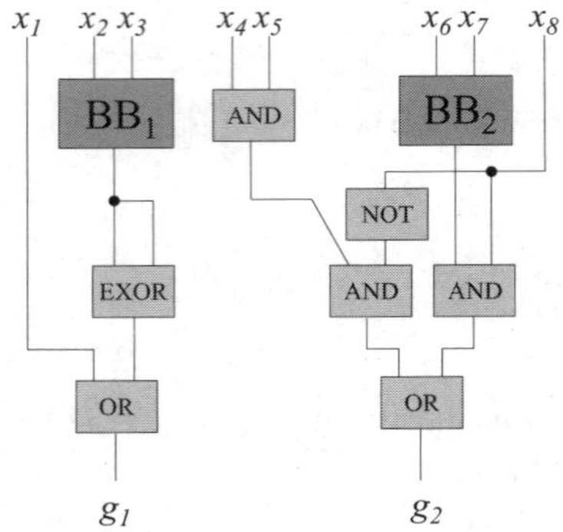

Figure 7.6. Disadvantage of symbolic Z-simulation.

Figure 7.6 shows an example for such a situation. The partial implementation shown does not fulfill the specification of Figure 7.1a, i.e., there is no implementation for the black boxes which leads to a correct overall implementation. However the approach of the previous section always computes X at the output of the EXOR-gate, since both inputs of the $EXOR$ gate are X (cf. Figure 7.5). That is, $g_1 = x_1 + Z$. Therefore, the first primary output is X, if $x_1 = 0$, and 1, if $x_1 = 1$. Since the first output of the specification is 1 as well, if $x_1 = 1$, no error can be detected at the first output. Moreover it is easy to see that the partial implementation of the second output is also correct (replace BB_2 by an OR-gate). So the method of the previous section cannot detect an error in this partial implementation.
If we have a closer look at the partial implementation, we can see, that the first output does not depend on the output of black box BB_1 as the output of the $EXOR$-gate whose inputs are connected to the output of BB_1 is 0 independently of the output of BB_1. The input vector $(0, 0, 0, 1, 1, 0, 0, 0)$ leads to output $(0, 1)$ which is different from the output $(1, 1)$ of the specification. The reason why the error could not be detected by symbolic Z–simulation lies in the fact that the values X at the inputs of the $EXOR$-gate are not accurately processed. The simple X-propagation does not take into account that the X-information comes from the same output of black box BB_1 (resulting in value 0 at the output).
To consider the origin of X–information it is not enough to introduce one variable Z for all black box outputs. Instead of that we have to introduce different

variables Z_i for each black box output i. After the inputs of the circuit have been associated with unique BDD variables a *conventional* symbolic simulation is performed. That is, the ROBDDs of the Boolean functions realized by the circuit under consideration are computed as described in Chapter 3 Section 3.2 on page 38. Now, the result for primary output j of the circuit is a function g_j which depends on the primary input variables $x_1, \ldots, x_n$ and the l variables $Z_1, \ldots, Z_l$ for the l outputs of the black boxes. Once again, let us illustrate the approach with the circuit already used for the illustration of the Z-simulation procedure (see Figure 7.4). In Figure 7.7, we have tagged the signal lines with the g_i's computed by the new approach, which is called Z_i*-simulation*.

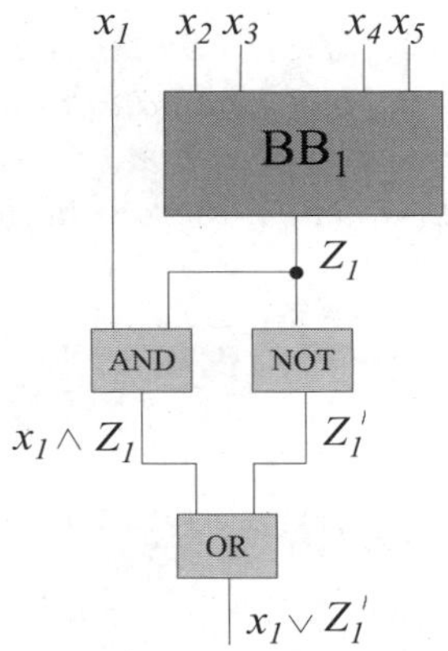

Figure 7.7. Illustration of Z_i-simulation

Conventional symbolic simulation of the circuit shown in Figure 7.6 with output variables Z_1 for BB_1 and Z_2 for BB_2, respectively, results in

$$g_1 = x_1$$

for the first output, which proves that the partial implementation is faulty, and

$$g_2 = (x_8{}' \cdot (x_4 \cdot x_5)) + (x_8 \cdot Z_2)$$

for the second output.

Now, let us go into details.

2.1 Local check

As in the case of Z–simulation we consider complete cofactors of implementation and specification with respect to all primary input variables. Here we call a cofactor *complete*, if it is a cofactor with respect to all *primary* input variables. If a complete cofactor of some output function of the partial implementation is

0 (**1**), this means that the output value is 0 (1) independently of the functionality of the black boxes. If for the same output function of the specification this cofactor is **1** (**0**), then we have found an error in the partial implementation. The only difference of this method compared to Z–simulation is the fact that the effect of the unknown values at the outputs of black boxes is evaluated more accurately.

A check whether there is a distinguishing vector for a primary output j of the partial implementation and the specification can be done accordingly to the following lemma. It is called *local check*, since it is done for each output separately.

LEMMA 7.3 (LOCAL CHECK) *Let g_j be the function of output j after symbolic Z_i–simulation for a partial implementation with primary inputs $x_1, \ldots, x_n$ and l outputs of black boxes with corresponding variables $Z_1, \ldots, Z_l$. Let f_j be output j of the specification. Then, there is no input vector $(\alpha_1, \ldots, \alpha_n) \in \{0,1\}^n$ with*

$$(g_j)_{[x_1,\ldots,x_n]=(\alpha_1,\ldots,\alpha_n)} = 1 \text{ and } f_j(\alpha_1, \ldots, \alpha_n) = 0$$

if and only if

$$((\forall Z_1 : (\forall Z_2 : (\ldots(\forall Z_l : g_j)))) \Rightarrow f_j) = 1$$

and there is no input vector $(\alpha_1, \ldots, \alpha_n) \in \{0,1\}^n$ with

$$(g_j)_{[x_1,\ldots,x_n]=(\alpha_1,\ldots,\alpha_n)} = 0 \text{ and } f_j(\alpha_1, \ldots, \alpha_n) = 1$$

if and only if

$$((\forall Z_1 : (\forall Z_2 : (\ldots(\forall Z_l : g_j')))) \Rightarrow f_j') = 1$$

PROOF: We first prove the if-part of the first statement.
Let us assume that

$$((\forall Z_1 : (\forall Z_2 : (\ldots(\forall Z_l : g_j)))) \Rightarrow f_j) \neq 1,$$

i.e., there is an assignment $(\alpha_1, \ldots, \alpha_n) \in \{0,1\}^n$ with

$$((\forall Z_1 : (\forall Z_2 : (\ldots(\forall Z_l : g_j)))) \Rightarrow f_j)(\alpha_1, \ldots, \alpha_n) = 0$$

which is equivalent to

$$(\forall Z_1 : (\forall Z_2 : (\ldots(\forall Z_l : g_j))))(\alpha_1, \ldots, \alpha_n) = 1 \text{ and } f_j(\alpha_1, \ldots, \alpha_n) = 0.$$

The equation $(\forall Z_1 : (\forall Z_2 : (\ldots(\forall Z_l : g_j))))(\alpha_1, \ldots, \alpha_n) = 1$ implies that for all assignments $(\delta_1, \ldots, \delta_l)$ to the variables $Z_1, \ldots, Z_l$ the equation

$$g_j(\alpha_1, \ldots, \alpha_n, \delta_1, \ldots, \delta_l) = 1$$

holds. Thus,

$$(g_j)_{[x_1,\ldots,x_n]=(\alpha_1,\ldots,\alpha_n)} = 1$$

holds, too. This closes the proof of the if-part of the statement.

Now, suppose that

$$((\forall Z_1 : (\forall Z_2 : (\ldots (\forall Z_l : g_j)))) \Rightarrow f_j) = 1$$

holds. Then we have to consider two cases for an arbitrary input vector $(\alpha_1, \ldots, \alpha_n) \in \{0,1\}^n$:

- If $(\forall Z_1 : (\forall Z_2 : (\ldots (\forall Z_l : g_j))))(\alpha_1, \ldots, \alpha_n) = 0$, then there is an assignment $(\delta_1, \ldots, \delta_l)$ to the variables $Z_1, \ldots, Z_l$ with

  $$g_j(\alpha_1, \ldots, \alpha_n, \delta_1, \ldots, \delta_l) = 0.$$

  This proves

  $$(g_j)_{[x_1,\ldots,x_n]=(\alpha_1,\ldots,\alpha_n)} \neq 1.$$

- If $(\forall Z_1 : (\forall Z_2 : (\ldots (\forall Z_l : g_j))))(\alpha_1, \ldots, \alpha_n) = 1$, then

  $$f_j(\alpha_1, \ldots, \alpha_n) = 1$$

  can be concluded.

Thus the first statement holds. Statement 2 can be proven in an analogous manner. ∎

2.2 Output exact check

The local check based on Z_i–simulation is more exact than Z–simulation. However implications between different outputs are not taken into account. In order to obtain an even more accurate check, a more "global" viewpoint has to be used. Figure 7.8 illustrates the difficulty with the local check approach. For the first output the only possibility to fulfill the specification of Figure 7.1a is to replace BB_1 by the function $x_4 \cdot x_5$. However, for the second output the only possibility to fulfill the specification is to replace BB_1 by $(x_4 \cdot x_5)'$. This implies that the partial implementation of Figure 7.8 is incorrect. In spite of that, the error cannot be detected by the local check procedure, since it is done for each output separately.

To detect errors of this type we have to compute "local conditions" for each output, which guarantee correctness for the single outputs, and then, we have to

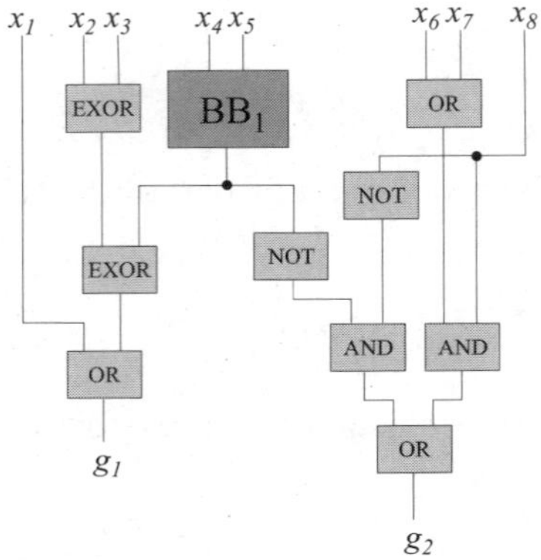

Figure 7.8. Disadvantage of the local check procedure.

combine the local conditions in order to check whether these local conditions can be fulfilled at the same time for all outputs.
The local conditions are computed based on the following considerations: To obtain a correct implementation, for each primary output j and each assignment $(\alpha_1, \dots, \alpha_n)$ to the primary inputs an assignment $(\delta_1, \dots, \delta_l)$ to the black box outputs has to be chosen such that

$$g_j(\alpha_1, \dots, \alpha_n, \delta_1, \dots, \delta_l) \quad \text{and} \quad f_j(\alpha_1, \dots, \alpha_n)$$

are identical, i.e.,

$$\left((g_j)_{[x_1,\dots,x_n]=(\alpha_1,\dots,\alpha_n)} \Leftrightarrow (f_j)_{[x_1,\dots,x_n]=(\alpha_1,\dots,\alpha_n)}\right)(\delta_1, \dots, \delta_l) = 1.$$

The relation between assignments $(\alpha_1, \dots, \alpha_n)$ to the primary inputs and assignments $(\delta_1, \dots, \delta_l)$ to the outputs of the black boxes, which are necessary to fulfill the specification, can be expressed by the characteristic function $cond_j$ defined by

$$\sum_{(\alpha_1,\dots,\alpha_n)\in\{0,1\}^n} (x_1^{\alpha_1} \cdot \dots \cdot x_n^{\alpha_n}) \cdot (g_j \Leftrightarrow f_j)_{[x_1,\dots,x_n]=(\alpha_1,\dots,\alpha_n)}$$

with $x_i^0 = x_i'$, $x_i^1 = x_i$, and $\sum$ denoting the iterated logical-or operator. The characteristic function $cond_j$ equals 1 for argument $(\alpha_1, \dots, \alpha_n, \delta_1, \dots, \delta_l)$, if and only if $g_j(\alpha_1, \dots, \alpha_n, \delta_1, \dots, \delta_l)$ and $f_j(\alpha_1, \dots, \alpha_n)$ are identical, i.e., if and only if the assignment of $(\delta_1, \dots, \delta_l)$ to the black box outputs leads to a correct output of the implementation for the given input $(\alpha_1, \dots, \alpha_n)$.
Applying the Boole-Shannon decomposition (see Chapter 2 Theorem 2.1) the expression for $cond_j$ can be rewritten as

$$cond_j = (g_j \Leftrightarrow f_j).$$

Thus, $cond_j$ can be computed using one ROBDD operation applied to the ROBDDs of g_j and f_j.

For a correct partial implementation all conditions $cond_1, \ldots, cond_m$ have to be true. If there is an input assignment $(\alpha_1, \ldots, \alpha_n)$ such that for all assignments $(\delta_1, \ldots, \delta_l)$ to the black box outputs at least one condition $cond_j$ is false, then the partial implementation cannot be used to obtain a correct final implementation. This leads us to a new, more accurate check, which we call *output exact check procedure*.

LEMMA 7.4 (OUTPUT EXACT CHECK) *If*

$$(\exists x_1 : \ldots (\exists x_n : (\forall Z_1 : \ldots (\forall Z_l : \sum_{j=1}^{m} (cond_j)')))) = 1$$

then the partial implementation does not fulfill its specification.

PROOF: The proof directly follows from the considerations made above. ∎

For illustration, consider the partial implementation of Figure 7.8 and the specification of Figure 7.1a. For input $(0,0,0,1,1,0,0,0)$, output g_1 of the implementation equals output f_1 of the specification only if the black box output is 1 for this input. This implies

$$cond_1(0,0,0,1,1,0,0,0,\delta_1) = 1 \Rightarrow \delta_1 = 1.$$

However, for input $(0,0,0,1,1,0,0,0)$ output g_2 of the implementation equals output f_2 only if the black box output is 0 for this input. This implies

$$cond_2(0,0,0,1,1,0,0,0,\delta_1) = 1 \Rightarrow \delta_1 = 0.$$

Thus, there is no assignment to δ_1 such that both $cond_1(0,0,0,1,1,0,0,0,\delta_1)$ and $cond_2(0,0,0,1,1,0,0,0,\delta_1)$ are equal to 1.
For the assignment $(0,0,0,1,1,0,0,0)$ to the primary inputs, it holds

$$\begin{aligned}
&(\forall Z_1 : (cond_1)' + (cond_2)')(0,0,0,1,1,0,0,0) \\
&= \left((cond_1)' + (cond_2)'\right)_{Z_1=1}(0,0,0,1,1,0,0,0) \\
&\quad \cdot \left((cond_1)' + (cond_2)'\right)_{Z_1=0}(0,0,0,1,1,0,0,0) \\
&= 1 \cdot 1 \\
&= 1.
\end{aligned}$$

That is, the output exact check reports an error as

$$(\exists x_1 : \ldots (\exists x_8 : (\forall Z_1 : (cond_1)' + (cond_2)'))) = \mathbf{1}.$$

2.3 Input exact check

The output exact check reports no error if and only if

$$(\forall x_1 : \ldots (\forall x_n : (\exists Z_1 : \ldots (\exists Z_l : \prod_{j=1}^{m} cond_j)))) = 1,$$

i.e., if and only if for each assignment $(\alpha_1, \ldots, \alpha_n)$ to the primary inputs there exists an assignment $(\delta_1, \ldots, \delta_l)$ to the black box outputs, such the conditions $cond_j$ for all outputs j are true, which means that $g_j(\alpha_1, \ldots, \alpha_n, \delta_1, \ldots, \delta_l)$ and $f_j(\alpha_1, \ldots, \alpha_n)$ are identical for all $1 \leq j \leq m$. Thus we can choose these values $\delta_1, \ldots, \delta_l$ to define the function values of the black box outputs under input $(\alpha_1, \ldots, \alpha_n)$. Thus, there is no error in the partial implementation, i.e., the black boxes can be replaced to obtain a correct final implementation, if the output exact check reports no error and *we are allowed to use all primary inputs as inputs of the black boxes*. However, this is not a realistic assumption. If we have fixed sets of input signals for the black boxes - actually they may be different from all primary inputs -, it is possible that the output exact check does not find all errors.

Figure 7.9 shows such a case. The partial implementation (for the specification

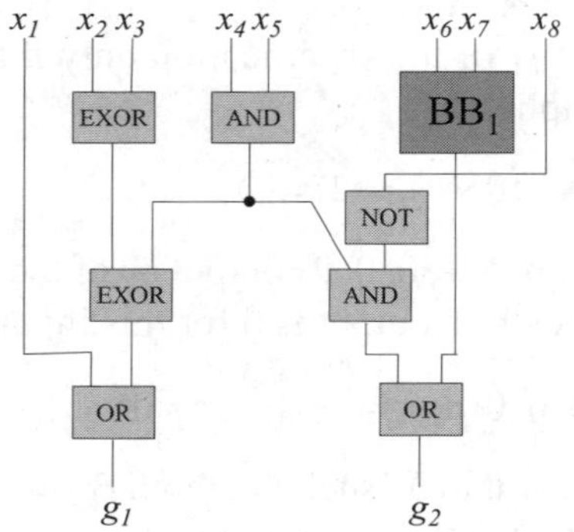

Figure 7.9. Disadvantage of output exact check

of Figure 7.1a) contains one black box BB_1. If the black box is replaced by $x_8 \cdot (x_6 + x_7)$, then implementation and specification are equivalent. It is easy to prove that a correct implementation for BB_1 must depend on input x_8. Just consider the assignment $x_4 = 0$, $x_5 = 0$, $x_6 = 1$, and $x_7 = 1$. For this assignment, output f_2 of the specification shown in Figure 7.1a is 0 if and only if $x_8 = 0$. The output g_2 of the partial implementation proposed in Figure 7.9 is 0 under this input assignment if and only if the output of the black box is 0. Thus, in order to obtain a correct implementation for black box BB_1, under the above assignment of the inputs x_4, x_5, x_6, and x_7, the output of black box BB_1 has to be 0 if and only if x_8 is 0. However, the inputs of black box BB_1 are only x_6 and x_7. Thus, there is no correct implementation for BB_1. The partial implementation is incorrect. But the output exact check does not report an error as

- $cond_1$ is a tautology and
- for all assignments of the primary inputs $x_1, \ldots, x_8$ there is an assignment to the output of black box BB_1, i.e., to Z_1, such that $cond_2$ evaluates to 1.

In the following, let us concentrate on defining a check which reflects this problem, too. We start with the special case that the partial implementation contains only one black box.

Before going into details, we have to define some abbreviations to simplify the notations.

- Let F be a Boolean formula over $X = \{x_1, \ldots, x_n\}$ and let $Y = \{y_1, \ldots, y_k\}$ be some subset of X. We abbreviate

  $$(\exists y_1 : \ldots (\exists y_k : F)) \text{ and } (\forall y_1 : \ldots (\forall y_k : F))$$

  by $(\exists Y : F)$ and $(\forall Y : F)$, respectively.
- The conjunction $\prod_{j=1}^{m} cond_j$ is abbreviated by $cond$. It can be interpreted as the characteristic function of a Boolean relation between assignments $(\alpha_1, \ldots, \alpha_n)$ to the primary input variables and assignments $(\delta_1, \ldots, \delta_l)$ to the outputs of the black boxes with

  $$cond(\alpha_1, \ldots, \alpha_n, \delta_1, \ldots, \delta_l) = 1$$

  if and only if $(\delta_1, \ldots, \delta_l)$ is a *legal* assignment to the outputs of the black boxes for primary input vector $(\alpha_1, \ldots, \alpha_n)$, i.e., if and only if all output values of the partial implementation with $(\alpha_1, \ldots, \alpha_n)$ assigned to the primary inputs and $(\delta_1, \ldots, \delta_l)$ assigned to the black box outputs are identical to the corresponding output values of the specification for assignment $(\alpha_1, \ldots, \alpha_n)$ to the primary inputs.

2.3.1 Partial implementations with only one black box

As mentioned, we first concentrate on partial implementations which contain only one black box, BB_1. We have to take into account that the inputs of this black box can be internal signals of the partial implementation and that not all primary inputs are connected to the inputs of BB_1. Let us assume that BB_1 has q input signal lines $i_1, \ldots, i_q$ and p output signal lines $o_1, \ldots, o_p$. We denote

the set of the input signal lines and output signal lines of BB_1 by I and O, respectively.
Now, consider the Boolean function which computes the assignments of the black box inputs. There are q such functions $h_1, \ldots, h_q$. They only depend on primary input variables of X. The characteristic function of the Boolean relation for $h_1, \ldots, h_q$ is computed by

$$H(x_1, \ldots, x_n, i_1, \ldots, i_q) = \prod_{k=1}^{q} (i_k \Leftrightarrow h_k(x_1, \ldots, x_n)).$$

To simplify the notation, we use the shortened form

$$H(X, I) = \prod_{k=1}^{q} (i_k \Leftrightarrow h_k(X)).$$

in the following.

Based on the Boolean relation *cond* defined above, which is a Boolean relation between primary input assignments and output assignments of black box BB_1, we compute the characteristic function of a Boolean relation between input assignments and output assignments of black box BB_1. Formally, this Boolean relation which we call $COND$ is defined as

$$COND(I, O) = (\forall X : H'(X, I) + cond(X, O)).$$

$COND(\epsilon, \delta)$ computes 1 for an assignment $\epsilon = (\epsilon_1, \ldots, \epsilon_q)$ to the inputs of black box BB_1 and an assignment $\delta = (\delta_1, \ldots, \delta_p)$ to the outputs of black box BB_1 if and only if for all assignments $\alpha = (\alpha_1, \ldots, \alpha_n)$ to the primary inputs

- α and ϵ lead to a signal assignment, which is not consistent with the circuit of the partial implementation – this is checked by term $H'(X, I)$

  or

- δ is a legal output of black box BB_1 under input α, i.e., α and δ result in correct values at the primary outputs of the partial implementation. This is checked by term $cond(X, O)$.

That is, for $COND(\epsilon, \delta)$ to be 1, vector δ has to be a legal output of black box under input α whenever α and ϵ lead to a signal assignment, which is consistent with the circuit of the partial implementation.

Now, the following theorem, which gives us a necessary and sufficient condition for the correctness of partial implementations with exactly one black box, can be proven.

THEOREM 7.1 (ONE BLACK BOX INPUT EXACT CHECK)
There is a replacement of black box BB_1 by a completely specified Boolean function with input variables I leading to a correct overall implementation if and only if

$$(\forall I : (\exists O : COND(I, O))) = \mathbf{1}$$

holds.

PROOF: Firstly, let us assume that

$$(\forall I : (\exists O : COND(I, O))) = \mathbf{1}$$

holds, i.e., for all $\epsilon \in \{0,1\}^q$ there is at least one vector $\delta \in \{0,1\}^p$ such that $COND(\epsilon, \delta) = 1$. Let δ_ϵ be any such vector which is related with ϵ.
Now, consider $bb_1 \in \mathcal{B}_{q,p}$ defined by

$$\forall \epsilon \in \{0,1\}^p : bb_1(\epsilon) = \delta_\epsilon.$$

We prove that a replacement of black box BB_1 by bb_1 leads to an overall circuit which fulfills the specification. Assume that this is false, i.e., there is an assignment α to the input variables for which the output of the partial implementation after replacement of black box BB_1 is incorrect. Let $\epsilon \in \{0,1\}^q$ be the input vector and $\delta \in \{0,1\}^p$ the output vector of black box BB_1 (actually replaced by bb_1), which is obtained by simulation, when vector α is applied to the inputs of the implementation. Then, we can argue as follows:

- Since the output is incorrect, $cond(\alpha, \delta) = 0$ holds.
- Since $bb_1(\epsilon) = \delta$, i.e., $\delta = \delta_\epsilon$, $COND(\epsilon, \delta) = 1$ holds by the definition of bb_1.
- Since $COND(\epsilon, \delta) = 1$ and $cond(\alpha, \delta) = 0$, we can conclude by the definition of $COND$ that $H(\alpha, \epsilon) = 0$ which states that α and ϵ is an inconsistent assignment to the signals of the implementation. However, this contradicts the fact that ϵ and δ were determined by simulation of the implementation applied to the input vector α.

This closes the proof that the replacement of black box BB_1 by bb_1 leads to an overall circuit which fulfills the specification.

Now, let us prove the other direction. We assume that there is a replacement of BB_1 by a completely specified Boolean function $bb_1 \in \mathcal{B}_{q,p}$, which leads to a correct implementation. We have to prove that

$$\forall \epsilon \in \{0,1\}^q \exists \delta \in \{0,1\}^p : COND(\epsilon, \delta) = 1.$$

Let $\epsilon \in \{0,1\}^q$ be fixed. We show that $COND(\epsilon, \delta_\epsilon)$ with $\delta_\epsilon := bb_1(\epsilon)$ holds.

Let $\alpha \in \{0,1\}^n$ be an arbitrary assignment to the input variables of the implementation. We have to consider two different cases.

- If $cond(\alpha, \delta_\epsilon) = 1$,

  $$H'(\alpha, \epsilon) + cond(\alpha, \delta_\epsilon) = 1$$

  immediately follows.

- If $cond(\alpha, \delta_\epsilon) = 0$, then the output of the partial implementation with α assigned to the primary inputs and δ_ϵ assigned to the outputs of black box BB_1 is not identical to the corresponding output of the specification. As the replacement of BB_1 by bb_1 has led to a correct implementation of the specification, the assignment ϵ to the inputs of black box BB_1 is not generated by assigning α to the primary inputs, i.e., $H(\alpha, \epsilon) = 0$ holds. Thus,

  $$H'(\alpha, \epsilon) + cond(\alpha, \delta_\epsilon) = 1$$

  holds in this case, too.

Since α was chosen as an arbitrary assignment to the primary input variables, this shows that

$$\begin{aligned} COND(\epsilon, \delta_\epsilon) &= (\forall X : H'(X, \epsilon) + cond(X, \delta_\epsilon)) \\ &= 1 \end{aligned}$$

holds. ■

Note that, in general, the necessary and sufficient condition given in Theorem 7.1 can be checked efficiently.

2.3.2 Partial implementations with more than one black box

The condition given in Theorem 7.1 can be generalized to partial implementations with more than one black box. However, the corresponding decision problem is NP-complete. For detailed information, we refer to [129, 130].
Here we confine the considerations to a straightforward generalization of the condition presented in Theorem 7.1 which results in a check for partial implementations with more than one black box which is not exact, but at least as good as the output exact check.
In the following, we assume that we have b black boxes $BB_1, \ldots, BB_b$ which can have several outputs and inputs. The set of the input signals of BB_j and

the set of the output signals of BB_j are denoted by I_j and O_j, respectively. Let us assume that black box BB_j has q_j input signal lines and p_j output signal lines. Furthermore, we assume that the black boxes are topologically ordered. Without loss of generality, assume that $BB_1, \ldots, BB_b$ is a topological order of the black boxes, i.e., BB_1 is the first black box with respect to this order.

Now, consider the Boolean function which computes the assignments to the inputs of the black boxes. For black box BB_j, there are q_j such functions $h_1^{(j)}, \ldots, h_{q_j}^{(j)}$. Because of the topological order of the black boxes, they can depend at most on primary input variables of X and the output variables of $O_1, \ldots, O_{j-1}$ of $BB_1, \ldots, BB_{j-1}$. We denote the characteristic function of the Boolean relation for $h_1^{(j)}, \ldots, h_{q_j}^{(j)}$ by $H_j(X, O_1, \ldots, O_{j-1}, I_j)$.
Based on the Boolean relation $cond(X, O_1, \ldots, O_b)$, which is a Boolean relation between primary input assignments and legal output assignments of the black boxes $BB_1, \ldots, BB_b$, we compute the characteristic function

$$COND(I_1, \ldots, I_b, O_1, \ldots, O_b)$$

of a Boolean relation between input assignments and output assignments of black box BB_1 which is formally defined as

$$(\forall X : H_1{}'(X, I_1) + \ldots + H_b{}'(X, O_1, \ldots, O_{b-1}, I_b) + cond(X, O_1, \ldots, O_b)).$$

Now, the new check is given by

$$(\forall I_1 : (\exists O_1 : \ldots (\forall I_b : (\exists O_b : COND(I_1, \ldots, I_b, O_1, \ldots, O_b))))) = 1$$

and it can be proven that the check is at least as good as the output exact check

THEOREM 7.2

$$(\forall I_1 : (\exists O_1 : \ldots (\forall I_b : (\exists O_b : COND(I_1, \ldots, I_b, O_1, \ldots, O_b)))))$$
$$\Longrightarrow \quad (\forall X : (\exists O_1 : \ldots (\exists O_b : cond(X, O_1, \ldots, O_b))))$$

PROOF: Assume that

$$(\forall I_1 : (\exists O_1 : \ldots (\forall I_b : (\exists O_b : COND(I_1, \ldots, I_b, O_1, \ldots, O_b))))) = 1$$

holds. Obviously, this assumption is equivalent to the statement, that there are Boolean functions $f_{O_1}, \ldots, f_{O_b}$ with

$$f_{O_j} : \{0,1\}^{\sum_{k=1}^{j} q_k} \longrightarrow \{0,1\}^{p_j}$$

such that

$$(\forall I : (\forall X : CC(X, I_1, \ldots, I_b, f_{O_1}(I_1), \ldots, f_{O_b}(I_1, \ldots, I_b)))) = 1$$

with $I = I_1 \cup \ldots \cup I_b$ and

$$\begin{aligned} & CC(X, I_1, \ldots, I_b, O_1, \ldots, O_b) \\ = \ & H_1{}'(X, I_1) + \ldots + H_b{}'(X, O_1, \ldots, O_{b-1}, I_b) + cond(X, O_1, \ldots, O_b). \end{aligned}$$

In order to prove that for an arbitrary vector $\alpha \in \{0,1\}^n$ there exist vectors $\delta^{(1)} \in \{0,1\}^{q_1}, \ldots, \delta^{(b)} \in \{0,1\}^{q_b}$ such that

$$cond(\alpha, \delta^{(1)}, \ldots, \delta^{(b)}) = 1$$

holds, we make use of the topological order of the black boxes. We apply α to the primary inputs of the partial implementation. This way, we obtain an input assignment $\epsilon^{(1)} \in \{0,1\}^{q_1}$ of black box BB_1 with $H_1(\alpha, \epsilon^{(1)}) = 1$. Applying α to the primary inputs and $f_{O_1}(\epsilon^{(1)})$ to the outputs of black box BB_1 we obtain an input assignment $\epsilon^{(2)} \in \{0,1\}^{q_2}$ of black box BB_2 with $H_2(\alpha, f_{O_1}(\epsilon^1), \epsilon^{(2)}) = 1$. Continuing this procedure we obtain $\epsilon^{(3)}, \ldots, \epsilon^{(b)}$ with

$$\forall j \in \{1, \ldots, b\} : H_j(\alpha, f_{O_1}(\epsilon^1), \ldots, f_{O_{j-1}}(\epsilon^{(1)}, \ldots, \epsilon^{(j-1)}), \epsilon^{(j)}) = 1.$$

Using our assumption, we have to conclude that

$$cond(\alpha, f_{O_1}(\epsilon^{(1)}), \ldots, f_{O_b}(\epsilon^{(1)}, \ldots, \epsilon^{(b)})) = 1$$

and since vector α was arbitrarily chosen

$$(\forall X : (\exists O_1 : \ldots (\exists O_b : cond(X, O_1, \ldots, O_b)))).$$

■

Notes

1 The results of this chapter have been presented in [129] and [130]. The adaption of these papers for the book was done in close cooperation with Paul Molitor.

Chapter 8

PERMUTATION INDEPENDENT BOOLEAN COMPARISON

A problem that occurs in equivalence checking of combinational circuits is that most verifiers use syntactical name matching to determine corresponding input variables and corresponding output variables. This is one of the most important reasons for false negative results. During synthesis huge hierarchical designs lead to long hierarchical names for signal lines. Since most backend tools have severe restrictions on name lengths, the names are shortened accordingly. In particular, if VHDL generics are involved, then the names may be changed such that not even a human is able to establish this correspondence without a detailed analysis of the design [118]. On the other hand, looking at equivalence checking methods it is obvious that things are much easier when the correspondence between the primary inputs and the primary outputs of the two circuits which have to be compared has been known. For example, before applying equivalence checking methods based on canonical representations it is even necessary to restore this correspondence since ROBDDs und *BMDs are canonical forms for a fixed variable order only.

Establishing the correspondence between the outputs is not a problem if the correspondence between the inputs has already been restored. Just sort the decision diagram pointers of the single-output Boolean functions of the specification and the decision diagram pointers of the single-output Boolean functions of the synthesized circuit with respect to the same fixed variable order and compare the two lists. Thus, only the correspondence between the input variables has to be established before checking equivalence.

This correspondence problem, which is called *permutation independent Boolean comparison problem*, is a particular *Boolean matching* problem. Usually, Boolean matching is applied to *library binding* and tackles the problem of recognizing whether a portion of a multiple-level logic circuit can be implemented by a cell of the cell library of a given technology [7]. This classification problem has already been investigated for a long time. Following the work of Dertouzos [39] and Lechner [87] it has been shown that spectral methods yield powerful classification techniques for functions. While this theory is fascinating, its implementation seems to be impractical for verification problems because of the exponential size of the matrices used by these methods. For more details on spectral techniques in logic design, we refer to the book of Hurst, Miller, and Muzio [65].

In contrast to library binding, the verification problem requires Boolean matching algorithms which are able to handle m-ary Boolean functions with a large number of input variables and need not to deal with the change of polarities, i.e., inversion of variables. Permutation independent Boolean comparison has been tackled by several researchers in the last few years. Mailhot and De Micheli [36, 91, 92] were the first to propose a method for library binding which is also applicable to permutation independent Boolean comparison and which was not a brute force method. In order to prune down equivalence checks, they used filters which compute properties of the input variables of the Boolean functions under comparison which represent necessary conditions for matching and which can easily be verified. In the case of completely specified Boolean functions, they take advantage of two facts to reduce the number of input permutations to be tested, namely that any permissible input permutation has to assign an unate (binate) variable (see Definition 8.4 on page 177) of the synthesized circuit to an unate (binate) variable of the specification and that interchangeable inputs of the synthesized circuit have to be interchangeable in the specification as well. The first condition implies that the Boolean functions which have to be compared must have the same number of unate (and binate) variables to have a match. The second condition implies that the synthesized circuit has to have a maximal set of pairwise interchangeable variables of size k whenever the specification contains such a maximal set of pairwise interchangeable variables of size k. This idea of filters has been extended, for example by Mohnke et al. [105, 106], Wang et.al. [143, 144], and Lai et al. [85]. They are called *signatures* in the following.

A signature is a description of an input variable which is independent of the permutation of the inputs of a Boolean function f. So, it can be used to identify this variable independent of permutation, i.e., any possible correspondence between the input variables of *two* functions is restricted to a correspondence between variables with the same signature. So, if each variable of a function f had a *unique* signature, then there would be at most one possible correspondence to the variables of any other function. That is why the quality of any signature is characterized by its ability to be a unique identification of a variable and, of course, by its ability to be computed fast. The signatures that have been introduced in literature differ in terms of quality. Nevertheless, in general this concept is a promising one and successful in a large number of practical cases. However, this method is not complete. There is *no* signature which can uniquely identify *all* the variables of the investigated benchmark sets. On comparing the most successful signatures, it can be observed that those benchmarks which have variables that cannot be uniquely identified have been always the same using different signatures. In other words, there is a nearly constant set of benchmarks for which signatures cannot help to solve the permutation problem. Unfortunately, not just 2 or 3 variables of those benchmarks are not uniquely identified, but about 15 and more, so that the number of possible correspondences is still large [27, 106, 108]. Furthermore, this seems to be independent of the method used to solve that problem. In [119], a totally different method for permutation independent Boolean comparison which does not apply signatures has been used. However, even here, the same group of benchmarks causes problems [120]. Another observation is that those benchmarks with non-uniquely identified variables are not the benchmarks with the most number of inputs. So, from a statistical point of view, it can be conjectured that the quality of the used signatures is not the problem. There is no relationship between the number of input variables of a function and the ability of the signatures to distinguish between all these inputs. This is due to the fact that the variables that cannot be distinguished by signatures have special properties that make it impossible to distinguish between them: the Boolean function under consideration is "symmetric" in these variables.

Now let us go into details.

1. The problem

The *permutation independent Boolean comparison problem* is defined as follows: Given two m-ary Boolean functions $f = (f_1, \ldots, f_m) \in \mathcal{B}_{n,m}$ and $g = (g_1, \ldots, g_m) \in \mathcal{B}_{n,m}$. Find a permutation of the variables of f and a permutation of the single-output Boolean functions $f_1, \ldots, f_m$ of f such that f becomes equivalent to g. More formally,

DEFINITION 8.1 (PERMUTATION INDEPENDENT COMPARISON)
Given two m-ary Boolean functions $f, g \in \mathcal{B}_{n,m}$ over X_n, the permutation independent Boolean comparison problem is the problem of finding a permutation $\tilde{\tau} \in \mathcal{P}er(\mathbb{N}_n)$ and a permutation $\rho \in \mathcal{P}er(\mathbb{N}_m)$ such that

$$(f_{\rho(j)} \circ \tau) = g_j$$

holds for $\tau \in \mathcal{P}er(\{0,1\}^n)$ defined by

$$\tau(x_1, \ldots, x_n) = (x_{\tilde{\tau}(1)}, \ldots, x_{\tilde{\tau}(n)}).$$

The m-ary Boolean functions f and g are said to be permutation equivalent or isomorphic if such permutations $\tilde{\tau}$ and ρ exist.

Note that $\tilde{\tau}$ which is a permutation of $\mathbb{N}_n$ induces a permutation $\overline{\tau} \in \mathcal{P}er(X_n)$ defined by

$$\forall x_i \in X_n : \overline{\tau}(x_i) = x_{\tilde{\tau}(i)},$$

and vice versa. For the sake of simplicity, we use $\tilde{\tau}$, τ, and $\overline{\tau}$ interchangeably and use the notation τ for each of the three functions, in the following. τ and ρ are called *input permutation* and *output permutation*, respectively.

The corresponding decision problem as to whether two Boolean functions are isomorphic is called *isomorphism problem* in computation theory [138]. If Boolean formulae are given for the two Boolean functions, the isomorphism problem is known to be NP-hard. The complexity of the isomorphism problem for ROBDDs is still unknown, even for single-output Boolean functions. On the one hand, no polynomial algorithm solving this problem is known. On the other hand, the problem has not been proven to be NP-hard. The only fact known about the problem is that it is in NP and that it is not NP-hard unless the polynomial hierarchy collapses [139].

The following sections follow [105, 108, 109].

2. Signatures

The best way to tackle the permutation independent Boolean comparison problem is to use signatures. We distinguish between signatures for inputs and signatures for outputs. In the following, let U be some ordered set.

DEFINITION 8.2 (INPUT SIGNATURE) *A function*

$$S_{in} : (\bigcup_{m \in \mathbb{N}} \mathcal{B}_{n,m}) \times \{x_1, \ldots, x_n\} \to U$$

is an input signature function if and only if the equation

$$S_{in}(f_1, \ldots, f_m, x_i) = S_{in}(f_{\rho(1)} \circ \tau, \ldots, f_{\rho(m)} \circ \tau, x_{\tau(i)})$$

holds for all $m \in \mathbb{N}$, for all single-output Boolean functions $f_1, \ldots, f_m \in \mathcal{B}_n$, for all input permutations τ, and for all output permutations ρ. We call the value $S_{in}(f, x_i)$ with $f = (f_1, \ldots, f_m)$ a signature of variable x_i with respect to the m-ary Boolean function f. S_n denotes the set of input signature functions with domain $(\cup_{m \in \mathbb{N}} \mathcal{B}_{n,m}) \times \{x_1, \ldots, x_n\}$, in the following.

DEFINITION 8.3 (OUTPUT SIGNATURE) *A function $S_{out} : \mathcal{B}_n \to U$ is called output signature function if and only if the equation*

$$S_{out}(f_j) = S_{out}(f_j \circ \tau)$$

holds for all single-output Boolean functions $f_j \in \mathcal{B}_n$ and for all input permutations τ.

As already mentioned, the correspondence between the outputs of the specification and the outputs of the synthesized circuit can be efficiently established if the correspondence between the input variables is already reconstructed. Thus, at a first glance, using output signatures does not help to solve our correspondence problem. However, knowledge about some output correspondences makes the input correspondence problem easier to solve since input signatures with respect to identified outputs can be used for restoring the correspondence between the input variables. We will illustrate this statement at the end of this section.

To illustrate the two definitions let us look at an example. Let

$$f = (f_1, f_2, f_3, f_4) \in \mathcal{B}_{3,4}$$

and

$$g = (g_1, g_2, g_3, g_4) \in \mathcal{B}_{3,4}$$

be two 4-ary Boolean functions defined by

$$\begin{aligned} f_1(x_1, x_2, x_3) &= (x_1 \cdot x_2') + x_3, \\ f_2(x_1, x_2, x_3) &= x_2 \cdot x_3, \\ f_3(x_1, x_2, x_3) &= x_1 \cdot x_3, \\ f_4(x_1, x_2, x_3) &= x_2 \end{aligned}$$

and

$$\begin{aligned} g_1(x_1, x_2, x_3) &= x_1 \cdot x_2, \\ g_2(x_1, x_2, x_3) &= (x_3 \cdot x_1') + x_2, \\ g_3(x_1, x_2, x_3) &= x_3 \cdot x_2, \\ g_4(x_1, x_2, x_3) &= x_1. \end{aligned}$$

For these two 4-ary Boolean functions it is easy to prove that they are isomorphic. Take function $\tau : \{1, 2, 3\} \to \{1, 2, 3\}$ defined by

$$\tau(1) = 3,\ \tau(2) = 1, \text{ and } \tau(3) = 2$$

as input permutation and the function

$$\rho : \{1, 2, 3, 4\} \to \{1, 2, 3, 4\}$$

defined by

$$\begin{aligned} \rho(1) &= 2, \qquad & \rho(3) &= 3, \\ \rho(2) &= 1, \qquad & \rho(4) &= 4 \end{aligned}$$

as output permutation. The equations

$$\begin{aligned}(f_{\rho(1)} \circ \tau)(x_1, x_2, x_3) &= f_2(\tau(x_1, x_2, x_3)) \\ &= f_2(x_{\tau(1)}, x_{\tau(2)}, x_{\tau(3)}) \\ &= f_2(x_3, x_1, x_2) \\ &= x_1 \cdot x_2 \\ &= g_1(x_1, x_2, x_3),\end{aligned}$$

$$\begin{aligned}(f_{\rho(2)} \circ \tau)(x_1, x_2, x_3) &= g_2(x_1, x_2, x_3), \\ (f_{\rho(3)} \circ \tau)(x_1, x_2, x_3) &= g_3(x_1, x_2, x_3), \\ (f_{\rho(4)} \circ \tau)(x_1, x_2, x_3) &= g_4(x_1, x_2, x_3)\end{aligned}$$

hold. Thus, for any input signature function S, the input signatures of x_2, x_3, and x_1 with respect to f are equal to the input signatures of x_1, x_2, and x_3, respectively, with respect to g.

In order to illustrate the signature based approach for solving the input correspondence problem, let us consider two very simple signature functions, namely the input signature function in_dep and the output signature function out_dep. The input signature $in\_dep(f, x_i)$ of a variable x_i with respect to a m-ary Boolean function $f = (f_1, \ldots, f_m)$ is the number of the single-output Boolean functions f_j of f that essentially depend on x_i, i.e.,

$$in\_dep(f_1, \ldots, f_m, x_i) = |\{f_j : \ f_j \text{ essentially depends on } x_i\}|.$$

A Boolean function f_j essentially depends on variable x_i if there exists an assignment $(\alpha_1, \ldots, \alpha_n) \in \{0,1\}^n$ of the variables of f_j, such that

$$f_j(\alpha_1, \ldots, \alpha_{i-1}, \alpha_i, \alpha_{i+1} \ldots, \alpha_n) \neq f_j(\alpha_1, \ldots, \alpha_{i-1}, \alpha_i{}', \alpha_{i+1}, \ldots, \alpha_n)$$

holds. The output signature $out\_dep(f_j)$ of a single-output Boolean function f_j is the number of variables which the Boolean function f_j depends on, i.e.,

$$out\_dep(f_j) = |\{x_i : \ f_j \text{ essentially depends on } x_i\}|.$$

Now, let us consider the two 4-ary Boolean functions $f = (f_1, f_2, f_3, f_4)$ and $g = (g_1, g_2, g_3, g_4)$ from above. The signature $in\_dep(f, x_1)$ of x_1 with respect to f is 2 because only f_1 and f_3 essentially depend on variable x_1. The signatures of x_2 and x_3 with respect to f are 3:

$$\begin{aligned}in\_dep(f, x_1) &= 2, \\ in\_dep(f, x_2) &= 3, \\ in\_dep(f, x_3) &= 3.\end{aligned}$$

Computing the input signatures with respect to g results in

$$\begin{aligned} in\_dep(g, x_1) &= 3, \\ in\_dep(g, x_2) &= 3, \\ in\_dep(g, x_3) &= 2. \end{aligned}$$

Thus, based on these input signatures, any input permutation τ for which f and g are equivalent has to satisfy the equation $\tau(1) = 3$. However, we cannot distinguish between variable x_2 and x_3 of f yet, since they have the same input signature with respect to f. Thus, $\tau(2)$ could either be 1 or 2.
Now, apply the output signature *out_dep* to the Boolean functions f_1, f_2, f_3, f_4, g_1, g_2, g_3, and g_4. We obtain

$$\begin{aligned} out\_dep(f_1) &= 3, & out\_dep(g_1) &= 2, \\ out\_dep(f_2) &= 2, & out\_dep(g_2) &= 3, \\ out\_dep(f_3) &= 2, & out\_dep(g_3) &= 2, \\ out\_dep(f_4) &= 1, & out\_dep(g_4) &= 1. \end{aligned}$$

Thus, any output permutation ρ for which f and g are equivalent has to satisfy the equations $\rho(2) = 1$ and $\rho(4) = 4$. Consequently, the Boolean function f_4 can correspond to the Boolean function g_4 only and we can use input signatures with respect to f_4 and g_4, respectively, in order to restore the correspondence between the input variables not yet uniquely identified. These input signatures are given by

$$\begin{aligned} in\_dep(f_4, x_2) &= 1, & in\_dep(g_4, x_1) &= 1, \\ in\_dep(f_4, x_3) &= 0, & in\_dep(g_4, x_2) &= 0. \end{aligned}$$

With this, all input variables of f and g are uniquely identified. There is only one input permutation τ for which the 4-ary Boolean functions f and g can be equivalent. The input permutation τ is defined by

$$\tau(1) = 3,\ \tau(2) = 1, \text{ and } \tau(3) = 2.$$

In general, an input signature function S partitions the set X_n of the variables of an m-ary Boolean function $f = (f_1, \ldots, f_m) \in \mathcal{B}_{n,m}$ into disjoint subsets

$$X(S, f, u) = \{x_i : x_i \in X_n \text{ and } S(f, x_i) = u\}.$$

Let

$$\Sigma(S, f) = \{(u, |X(S, f, u)|) : |X(S, f, u)| \neq 0\}$$

be the set that characterizes the nonempty subsets. Then the following lemma holds.

LEMMA 8.1 *If f and g are isomorphic then $\Sigma(S, f)$ is equal to $\Sigma(S, g)$.*

The subsets $X(S, f, u)$ whose sizes are greater than 1 play an important role in the following. Let us call them *aliasing groups*.

LEMMA 8.2 *For any m-ary Boolean function f, at most*

$$\prod_{u \in U,\ |X(S,f,u)| \neq 0} |X(S, f, u)|!$$

input permutations have to be applied to solve the permutation independent comparison problem for f and any other m-ary Boolean function g.

Thus, the problem of permutation independent Boolean comparison with respect to combinational formal logic verification has been reduced to the problem of uniquely distinguishing the input variables of an m-ary Boolean function by applying input and output signature functions. In practice, it even is sufficient to reduce the number of aliasing groups to a few with sizes not larger than five to six variables each. In this case all remaining variable correspondences between the two functions f and g can be checked in moderate time.

3. Examples of signatures

In the following, we present input and output signature functions that can be used as filters during permutation independent Boolean comparison of m-ary Boolean functions. In particular, we deal with the ROBDD based computation of the proposed signatures. For this, let f be an m-ary Boolean function consisting of m single-output Boolean functions $f_1, \ldots, f_m \in \mathcal{B}_n$ represented by an ROBDD G. Remember that $|G|$ denotes the size of the ROBDD G.

3.1 Input signature functions

3.1.1 Using the support of Boolean functions

A first input signature function denoted by in_dep has already been presented in the last section. It computes the number of single-output Boolean functions of f that essentially depend on x_i, i.e., the size of the support of f. This data can efficiently be computed in time $\mathcal{O}(n \cdot |G|)$ by a bottom-up pass over G. For any node w of G with label x_i and successors w_0 and w_1, the support $support(\phi(w))$ of $\phi(w)$, which is the Boolean function represented by node w (see Definition 3.7 on page 32), is given by

$$support(\phi(w)) = support(\phi(w_0)) \cup support(\phi(w_1)) \cup \{x_i\}.$$

After this bottom-up pass, the signature $in\_dep(f, x_i)$ can be computed by the formula

$$in\_dep(f, x_i) = |\{f_j : \ x_i \in support(f_j)\}|.$$

3.1.2 Using the unateness of Boolean functions

The next input signature function which has been proposed by Mailhot and De Micheli [92] gives information about whether a variable is unate or binate.

DEFINITION 8.4 (UNATE/BINATE VARIABLE)

- *A variable x_i is positive unate with respect to f, if for all $j \in \mathbb{N}_m$ the negative cofactor $(f_j)_{x_i=0}$ is smaller (see Definition 2.9 on page 15) than the positive cofactor $(f_j)_{x_i=1}$.*
- *A variable x_i is negative unate with respect to f, if for all $j \in \mathbb{N}_m$ the positive cofactor $(f_j)_{x_i=1}$ is smaller than the negative cofactor $(f_j)_{x_i=0}$.*
- *A variable x_i is binate with respect to f, if x_i is neither positive unate nor negative unate with respect to f.*

Formally, the input signature function $unate$ can be defined by

$$unate(f, x_i) = \begin{cases} -1, & \text{if } x_i \text{ is negative unate with respect to } f \\ 1, & \text{if } x_i \text{ is positive unate with respect to } f \\ 0, & \text{if } x_i \text{ is binate with respect to } f. \end{cases}$$

This input signature function can be made more sophisticated by computing the number of single-output Boolean functions in which the variable x_i is positive unate and the number of single-output Boolean functions in which x_i is negative unate, i.e., by using the input signature function

$$UNATE : (\bigcup_{m \in \mathbb{N}} \mathcal{B}_{n,m}) \times \{x_1, \ldots, x_n\} \to \mathbb{N}_0 \times \mathbb{N}_0$$

with

$$\begin{aligned} & UNATE(f, x_i) \\ & = \; (\, |\{f_j : \; x_i \text{ is positive unate with respect to } f_j\}|, \\ & \qquad |\{f_j : \; x_i \text{ is negative unate with respect to } f_j\}| \,). \end{aligned}$$

Checking whether a single-output Boolean function f_j is positive unate or negative unate can be done by using the fact that the set $\mathcal{B}_n$ of the completely specified Boolean functions with n inputs forms a partially ordered set. Just generate the ROBDDs of the cofactors $(f_j)_{x_i=0}$ and $(f_j)_{x_i=1}$ and check whether

$$(f_j)_{x_i=0} + (f_j)_{x_i=1} = (f_j)_{x_i=1}$$

or

$$(f_j)_{x_i=0} + (f_j)_{x_i=1} = (f_j)_{x_i=0}$$

hold. This can be done in expected time $\mathcal{O}(|G|^2)$ (see Chapter 3). Thus the computation of the signature $UNATE(f, x_i)$ can be done in expected time $\mathcal{O}(m \cdot |G|^2)$.

3.1.3 Satisfy count signatures

A further class of input signature functions can be obtained by considering the satisfy counts of some particular Boolean sub-functions $h_j \in \mathcal{B}_n$ of the single-output Boolean functions f_j.

Cofactor satisfy count signature. A very simple input signature function which is based on the satisfy count is the cofactor satisfy count, which has been independently introduced by different researchers [27, 85, 106]. They take the positive cofactor $(f_j)_{x_i=1}$ for the Boolean function h_j from above. Formally, the cofactor satisfy count signature can be defined by

$$S_{cofactor}(f, x_i) = \{(k, m_{k,x_i}) : k \in \mathbb{N}_0 \text{ and } m_{k,x_i} > 0\}$$

where m_{k,x_i} is the number of single-output Boolean functions f_j of the m-ary Boolean function f with $|(f_j)_{x_i=1}|=k$. Thus, for all $k \in \mathbb{N}_0$, the cofactor satisfy count signature of variable x_i gives the number of single-output Boolean functions f_j whose positive cofactor $(f_j)_{x_i=1}$ has exactly k satisfying input assignments.

For the computation of the satisfy counts of $(f_1)_{x_i=1}, \ldots, (f_m)_{x_i=1}$, we can use the same bottom-up algorithm as described in Chapter 3 Section 3.2.5 with the only difference that at nodes w labelled with x_i the low-edge must not be considered and the computation step has to be substituted by

$$|\phi(w)| = 2 \cdot |\phi(w_1)| \cdot 2^{dist(w,w_1)-1}$$

w_1 denoting the high-successor of node w. Thus, the computation only performs $\mathcal{O}(|G|)$ operations (see Chapter 3 Section 3.2.5 on page 42).

Considering the satisfy counts of the negative cofactors $(f_1)_{x_i=0}, \ldots, (f_m)_{x_i=0}$ does not provide new information about the input variables because of the property

$$2 \cdot |f_j| = |(f_j)_{x_i=0}| + |(f_j)_{x_i=1}|.$$

Existential abstraction satisfy count signature. Another input signature function which is based on the satisfy count uses the satisfy counts of the *existential abstractions* $(\exists x_i : f_1), \ldots, (\exists x_i : f_m)$ of x_i with respect to the single-output Boolean functions $f_1, \ldots, f_m$ [105, 106]. Formally, this input signature function is defined by

$$S_{\exists}(f, x_i) = \{(k, p_{k,x_i}) : k \in \mathbb{N}_0 \text{ and } p_{k,x_i} > 0\},$$

where p_{k,x_i} is the number of single-output Boolean functions f_j of the m-ary Boolean function f with $|(\exists x_i : f_j)| = k$. The computation of the existential

abstraction satisfy count signature of a variable x_i with respect to f is a little bit more expensive than the computation of the signature $S_{cofactor}(f, x_i)$ because we have to construct the ROBDDs of $(\exists x_i : f_1), \ldots, (\exists x_i : f_m)$, which needs time $\mathcal{O}(m \cdot |G|^2)$.
Considering the satisfy counts of the universal abstractions $(\forall x_i : f_1), \ldots, (\forall x_i : f_m)$ or the satisfy counts of the Boolean differences $\partial f_1/\partial x_i, \ldots, \partial f_m/\partial x_i$ does not provide new information about the input variables because of the two properties

$$2 \cdot |f_j| = |(\exists x_i : f_j)| + |(\forall x_i : f_j)|$$

and

$$2 \cdot |f_j| = 2 \cdot |(\exists x_i : f_j)| - |\partial f_j/\partial x_i|.$$

Breakup signatures. A further class of input signature functions which are based on satisfy count as well considers the Hamming distances between the satisfying input assignments and a permutation independent origin o. The idea is to break the satisfy count of a Boolean function over n variables into $n+1$ special, permutation independent components. It has been proposed in [105, 106]. Let us explain the idea by first considering only one single-output function $f_j \in \mathcal{B}_n$. We take the vector $(0, \ldots, 0) \in \{0,1\}^n$ as permutation independent origin o. We are interested in the number of satisfying input assignments $\alpha \in \{0,1\}^n$ of f_j with Hamming distance d from the origin o. Formally, this can be defined by a function

$$\Psi : \mathcal{B}_n \to \mathbb{N}_0^{n+1}$$

with

$$\Psi(f_j) = (|f_j^0|, \ldots, |f_j^d|, \ldots, |f_j^n|),$$

and

$$f_j^d(\alpha) = \begin{cases} 1, & \text{if } f_j(\alpha) = 1 \text{ and } \sum_{i=1}^n \alpha_i = d \\ 0, & \text{otherwise.} \end{cases}$$

In practice two breakup signature functions have been proven to be useful:

- $\Psi((f_j)_{x_i=1})$ which uses the cofactor function of f_j with respect to a variable x_i and
- $\Psi(\exists x_i : f_j)$ which uses the existential abstraction of f_j with respect to a variable x_i.

As an example, let us consider function

$$f(x_1, x_2, x_3) = (x_1 \cdot x_2) + (x_1' \cdot x_3)$$

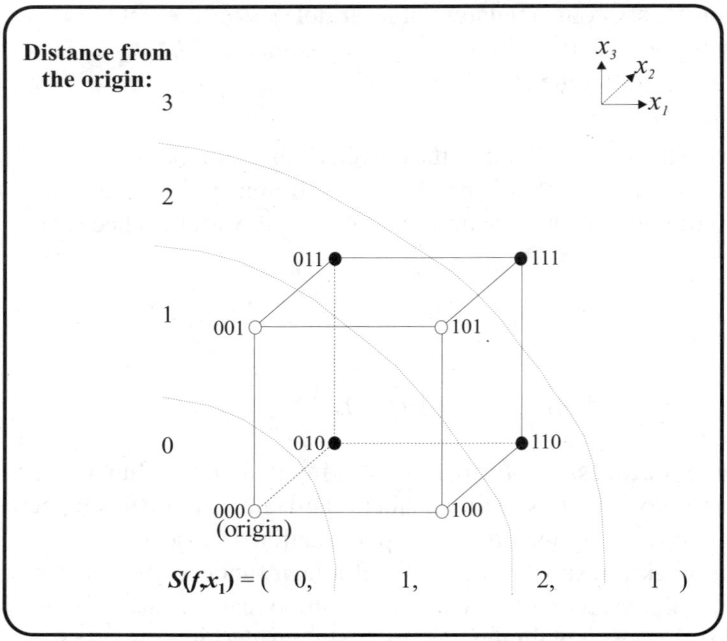

Figure 8.1. Breakup Signature for $f_{x_1=1} = x_2$ with $f = (x_1 \cdot x_2) + (x_1' \cdot x_3)$ and $o = (0, 0, 0)$

and its input variable x_1. Figure 8.1 shows a cube representation for the positive cofactor function $f_{x_1=1}$ which is x_2. Black dots indicate vertices where the cofactor function is equal to 1, white dots indicate the vertices where the cofactor function is equal to 0. With origin $o = (0, 0, 0)$ the vertices can be separated in four different, permutation independent levels as you can see in the figure. The dth level indicates the dth component in the breakup signature which is the number of black dots with distance d from the origin. This information is permutation independent since the origin o is permutation independent, the distance is permutation independent because it groups together all vertices with a constant number of variables equal to 1 independently of the order of these variables themselves, and the counting step itself is permutation independent because just the *number* of black dots in a level is of interest, not their *places* in this cube level.

The generalization of this idea to m-ary functions $f = (f_1, \ldots, f_m)$ is quite easy. It works with the same principle as before. For any $k, d \in \mathbb{N}_0$, compute the numbers of single-output Boolean functions of f whose positive cofactors (existential abstractions) with respect to x_i have exactly k satisfying input assignments with Hamming distance d from the origin o. Formally, the two breakup signature functions can be defined by

$$S^{breakup}_{cofactor}(f, x_i) = \{(k, d, q_{k,d,x_i}) : k, d \in \mathbb{N}_0 \text{ and } q_{k,d,x_i} > 0\}$$

where q_{k,d,x_i} is the number of single-output Boolean functions f_j of the m-ary Boolean function f with $\Psi((f_j)_{x_i=1})^d = k$ and

$$S^{breakup}_{\exists}(f, x_i) = \{(k, d, r_{k,d,x_i}) : k, d \in \mathbb{N}_0 \text{ and } r_{k,d,x_i} > 0\}$$

where r_{k,d,x_i} is the number of single-output Boolean functions f_j of the m-ary Boolean function f with $\Psi(\exists x_i : f_j)^d = k$. The computation of these two signatures of a variable x_i performs $\mathcal{O}(|G| \cdot n)$ and $\mathcal{O}(|G|^2 \cdot n)$ operations, respectively.

Cross signature. Schlichtmann, Brglez, and Schneider [127] proposed to use not only the satisfy counts of $(f_j)_{x_i=1}$ $(j = 1, \ldots, m)$ but also the satisfy counts of

$$(f_j)_{[x_i,x_1]=(1,0)}, \ldots, (f_j)_{[x_i,x_n]=(1,0)}$$

to uniquely identify variable x_i. More formally, the signature $S_{cr}(f_1, \ldots, f_m, x_i)$ of variable x_i with respect to $f_1, \ldots, f_m$ is given by the m ordered lists $L(f_j, x_i)$ of the satisfy counts

$$|(f_j)_{[x_i,x_1]=(1,0)}|, \ldots, |(f_j)_{[x_i,x_n]=(1,0)}|$$

for $j = 1, \ldots, m$. The lexicographically ordered list $L(f_1, \ldots, f_m, x_i)$ of the vectors $L(f_1, x_i), \ldots, L(f_m, x_i)$ is obviously a permutation independent information about variable x_i. A generalization of this cross signature function has been presented by Wang, Hwang, and Chen [144].

3.1.4 Function signatures

Function signatures of input variables with respect to an m-ary Boolean function $f = (f_1, \ldots, f_m)$ are not values or vector of values, but particular Boolean functions or vectors of Boolean functions. These signatures can be applied when some input variables have already been uniquely identified. The idea is to use sub-functions of the single-output Boolean functions $f_1, \ldots, f_m$ which only depend on input variables which have already been uniquely identified. Assume that the variables $x_{k+1}, \ldots, x_n$ have been uniquely identified. Then, the two Boolean functions

$$(f_j)_{[x_1,\ldots,x_{i-1},x_i,x_{i+1},\ldots,x_k]=(1,\ldots,1,0,1,\ldots,1)} \circ \tau_{can}$$

and

$$(f_j)_{[x_1,\ldots,x_{i-1},x_i,x_{i+1},\ldots,x_k]=(0,\ldots,0,1,0,\ldots,0)} \circ \tau_{can}$$

provide permutation independent information about the input variable x_i if τ_{can} is an input permutation that constructs a unique order of the uniquely identified input variables $x_{k+1}, \ldots, x_n$.

3.2 Output signature functions

Since output signature functions are based on input signature functions in most cases, we touch on this problem very briefly.

3.2.1 Using the support of a Boolean function

A first output signature function denoted by out_dep has already been proposed. It computes the number of variables which a single-output Boolean function f_j essentially depends on. The computation of this data can be done by depth-first-search on the ROBDD G_{f_j}. This process can be stopped when all n input variables have been encountered.

3.2.2 Satisfy count output signatures

Satisfy count output signatures can be directly developed by applying input signature functions as introduced in Section 3.1.3. The most important are

- the satisfy count $|f_j|$,
- the ordered vector of the cofactor satisfy counts $|(f_j)_{x_1=1}|, \ldots, |(f_j)_{x_n=1}|$,
- the breakup signature with respect to f_j.

3.2.3 Structure of symmetry sets

A more sophisticated output signature can be obtained by computing the maximal sets of pairwise interchangeable input variables of f_j.

DEFINITION 8.5 (SYMMETRIC BOOLEAN FUNCTION)
A single-output Boolean function $f_j \in \mathcal{B}_n$ is symmetric in the input variables x_i and x_k ($i < k$) if and only if

$$f_j(\alpha_1, \ldots, \alpha_i, \ldots, \alpha_k, \ldots, \alpha_n) = f_j(\alpha_1, \ldots, \alpha_k, \ldots, \alpha_i, \ldots, \alpha_n)$$

holds for all $\alpha \in \{0,1\}^n$. x_i and x_k are called pair of symmetric variables.

It is well known that symmetry as defined in Definition 8.5 leads to an equivalence relation on the set X_n of the input variables of f_j. The equivalence classes are the maximal subsets of pairwise interchangeable input variables. Obviously,

$$S_{symm}(f_j) = \{(k, t_k) : \ k \in \mathbb{N} \text{ and } t_k > 0\},$$

where t_k is the number of equivalence classes of size k, is a permutation independent information about f_j and represents an output signature function.

To compute the equivalence classes, we use the fact that a single-output Boolean function f_j is symmetric in x_i and x_j if and only if

$$(f_j)_{[x_i,x_j]=(1,0)} = (f_j)_{[x_i,x_j]=(0,1)}$$

holds. In a preprocessing step, we use filters to detect as many asymmetries as possible to prune the number of symmetry checks because each symmetry check has to construct ROBDDs for the cofactor functions, which is time consuming. Some of these filters use input signature functions. The idea behind it is that a single-output Boolean function f_j is not symmetric in the input variables x_i and x_j if x_i and x_j do not have the same input signature. We will discuss this statement in the next section. More details on efficient computation of maximal sets of pairwise interchangeable input variables can be found in [110, 132].

4. Limits of signatures

The main problem that arises in this paradigm is when different variables of a function f have the same signature, so that it is not possible to distinguish between these variables, i.e., there is no unique correspondence that can be established with the inputs of any other function. A group of such variables is called an *aliasing group*. Suppose there is just one aliasing group of inputs of a function f after applying certain signatures. If the size of this group is k, then there are still $k!$ correspondence possibilities to test between the inputs of f and the inputs of any other function g, in general (cf. Lemma 8.2). However, in certain special cases, aliasing is not dramatic. For instance, assume that the Boolean function under consideration is invariant under the permutation of the input variables $x_{i_1}, \ldots, x_{i_j}$, i.e., f is *symmetric* in the variables $\{x_{i_1}, \ldots, x_{i_j}\}$, then any order of $x_{i_1}, \ldots, x_{i_j}$ is fine for the purpose to test permutation equivalence although $x_{i_1}, \ldots, x_{i_j}$ are contained in the same aliasing group. Thus, whenever we know that f is symmetric in j variables of an aliasing group, the number of correspondences which have to be tested by Lemma 8.2 decreases by factor $j!$.

What can we say about the practical experiences with using signatures? When considering the benchmarks of the LGSynth91 [89] and the ESPRESSO benchmark set [18], approximately 92% of all benchmarks have a unique correspondence for their inputs, i.e., all their variables can be uniquely identified assuming that symmetric variables (as defined in Definition 8.5) are used during the identification process as just explained. For the 8% of benchmarks with aliasing, the number of possibilities for correspondence among inputs ranges from 4 to approximately 10^{23} after applying all signatures [105]. Considering the complete set of benchmarks, we can make the observation that there is no relationship between the number of input variables of a function and the ability of the signatures to distinguish between all these inputs. The problem seems to be that there are special properties that make it impossible to distinguish between those variables via signatures. Since signatures work well for about 92% of the benchmarks, there is no reason to reject the signature approach. What we need is a moderate solution for the other 8% as well. For circuits with just a few correspondence possibilities, the obvious solution of enumerating all possibilities works well. But what about the other circuits – for those, we cannot be satisfied with existing solutions. A further understanding of these cases is the focus of the remaining sections of this chapter. At first, we discuss a property of input variables which makes it impossible to distinguish between variables with aliasing with the help of signatures. This property has to do with symmetry — it is the so called $\mathcal{G}$–symmetry. We introduce $\mathcal{G}$–symmetry and explain why $\mathcal{G}$–symmetry avoids a unique identification of input variables with the help of signatures. To be able to explain the concepts more clearly we restrict our considerations to Boolean functions $f \in \mathcal{B}_n$ with one output only.

$\mathcal{G}$–symmetry can be defined as follows:

DEFINITION 8.6 ($\mathcal{G}$-SYMMETRY) *Consider a subgroup $\mathcal{G} \subseteq \mathcal{P}er(X_n)$ of permutations of the variables $\{x_1, \ldots, x_n\}$. A Boolean function $f \in \mathcal{B}_n$ is $\mathcal{G}$–symmetric if f keeps invariant under all input permutations τ in $\mathcal{G}$, i.e.,*

$$\forall \tau \in \mathcal{G} : \; f \circ \tau = f.$$

$\mathcal{G}$–symmetry was defined similar by Hotz in 1974 [63]. The simplest example for $\mathcal{G}$–symmetry is a Boolean function f that is symmetric in all input variables. Here, $\mathcal{G}$ is equal to the permutation group $\mathcal{P}er(X_n)$, and we say, f is $\mathcal{P}er(X_n)$–symmetric.

By definition, the group $\mathcal{G}$ may also be the group which contains the identity $\mathbf{1}$ only. So we can consider those Boolean functions with no $\mathcal{G}$–symmetry as functions that are $\mathcal{G}$–symmetric with respect to the set $\mathcal{G} = \{\mathbf{1}\}$.

In the following, let $\mathcal{G}(f) \subseteq \mathcal{P}er(X_n)$ be the set of all the input permutations under which the Boolean function f keeps invariant. Then, we can formulate the following properties:

LEMMA 8.3 *For each function $f \in \mathcal{B}_n$, set $\mathcal{G}(f)$ forms a group.*

PROOF: Let $\tau, \tau_1, \tau_2 \in \mathcal{G}(f)$. We need to prove that $\tau^{-1} \in \mathcal{G}(f)$ and $\tau_1 \circ \tau_2 \in \mathcal{G}(f)$ hold. Both statements follow by

$$\begin{aligned} f \circ \tau^{-1} &= (f \circ \tau) \circ \tau^{-1} \\ &= f \circ (\tau \circ \tau^{-1}) \\ &= f \end{aligned}$$

and

$$\begin{aligned} f \circ (\tau_1 \circ \tau_2) &= (f \circ \tau_1) \circ \tau_2 \\ &= f \circ \tau_2 \\ &= f. \end{aligned}$$

In other words, both the input permutation τ^{-1} and the input permutation $\tau_1 \circ \tau_2$ keep the function f invariant, thus $\tau^{-1} \in \mathcal{G}(f)$ and $\tau_1 \circ \tau_2 \in \mathcal{G}(f)$ hold. ▪

In order to understand the significance of $\mathcal{G}$–symmetry for the permutation equivalence problem, let us consider the relation $R_{\mathcal{G}}$ on the set of input variables, $X_n = [x_1, x_2, \ldots, x_n]$, which is defined by a group of permutations $\mathcal{G} \subseteq \mathcal{P}er(X_n)$. Formally, relation $R_{\mathcal{G}}$ is defined by

$$\forall\, x_i, x_j \in X_n : \;\; x_i R_{\mathcal{G}} x_j :\Longleftrightarrow \exists \tau \in \mathcal{G} : \tau(x_i) = x_j.$$

LEMMA 8.4 *The relation $R_{\mathcal{G}}$ is an equivalence relation.*

PROOF: This follows immediately from the property that $\mathcal{G}$ is a group. ▪

Let us denote with $\mathcal{A} = \{A_1, A_2, \ldots, A_k\}$ the set of equivalence classes with respect to $R_{\mathcal{G}}$ which partitions the input variables of X_n.

LEMMA 8.5 *For each Boolean function $f \in \mathcal{B}_n$ and its set of input variables X_n, the partition $\mathcal{A}$ of the input variables of f which corresponds to the equivalence relation $R_{\mathcal{G}(f)}$ is well–defined.*

Often it is enough to consider this partition $\mathcal{A}$ as an *unordered* set of subsets of the inputs. However, sometimes we have to look at it as an *ordered* set.

LEMMA 8.6 *Given a Boolean function $f \in \mathcal{B}_n$ and the well–defined partition $\mathcal{A} = \{A_1, A_2, \ldots, A_k\}$ of its input variables, there is a well–defined and ordered partition $\mathcal{A}^f = [A_1^f, A_2^f, \ldots, A_k^f]$ of the input variables with respect to f which is constructed as follows:*

1 $\mathcal{A}^f$ considered as unordered set of subsets is equal to partition $\mathcal{A}$.

2 The ordering step works as follows:

(a) Order the subsets by their size, such that $|A_i^f| \le |A_{i+1}^f|$ for all $i \in \mathbb{N}_k$.

(b) If some subsets have the same size, then use the elements of them to establish an ordering, such that $j < l$ holds if

- $|A_i^f| = |A_{i+1}^f|$,
- $j = \min\{h : x_h \in A_i^f\}$, *and*
- $l = \min\{h : x_h \in A_{i+1}^f\}$.

PROOF: The partition $\mathcal{A}^f$ contains the same elements as the original partition $\mathcal{A}$. So it is well–defined as an *unordered* set of subsets. Furthermore, the ordering process is well–defined. Step $2a$ orders those subsets uniquely that have different sizes. Step $2b$ orders the subsets with the same size uniquely. Note, that this second step depends on the order of the input variables of the function f. In other words, the order of the subsets in partition $\mathcal{A}^f$ is just unique with respect to the Boolean function f which has been considered. ▪

Now we consider signatures again. What can we say about the relationship between signatures and $\mathcal{G}$–symmetry? There are two properties that complete our picture about signatures. The first property is very important and can be formulated without further considerations.

THEOREM 8.1 *Let $\mathcal{G} \subseteq \mathcal{P}er(X_n)$ be the group of permutations of X_n which constructs partition $\mathcal{A}$. Consider any element A_l of $\mathcal{A}$, any $\mathcal{G}$–symmetric Boolean function f, and any signature function $s \in \mathcal{S}_n$. Then, for any $x_i, x_j \in A_l$ the equation $s(f, x_i) = s(f, x_j)$ holds.*

PROOF: [1] For all $x_i, x_j \in A_l$ there is a permutation $\tau \in \mathcal{G}$ such that $\tau(i) = j$. Consider any $x_i, x_j \in A_l$, a permutation $\tau \in \mathcal{G}$ such that $\tau(i) = j$, and any signature function s. Because of Definition 8.2, the equation

$$s(f, x_i) = s(f \circ \tau, \tau(x_i))$$

holds. Since f is $\mathcal{G}$–symmetric, we have

$$s(f \circ \tau, \tau(x_i)) = s(f, \tau(x_i)).$$

Together with the previous equation this results in

$$s(f, x_i) = s(f, x_j).$$

▪

This result now gives us a possible explanation for the trouble several benchmarks have with the signature approach. It is possible that the benchmarks with aliasing include $\mathcal{G}$–symmetric functions. Then there is no unique description by signatures for the input variables of these functions. Thus, it is futile to try to distinguish all the variables with additional signatures. But, what do we do in this case? Our response to this is that we do not really need to uniquely identify the variables in most cases. We can achieve our end goal of establishing permutation independent function equivalence by identifying the variables involved in the $\mathcal{G}$–symmetry and exploring the exact nature of the $\mathcal{G}$–symmetry. This is further explored in the next sections. Before we discuss it, let us develop the second property of signatures with respect to $\mathcal{G}$–symmetry. For this, we need to think about a universal signature function.

DEFINITION 8.7 (UNIVERSAL SIGNATURE) *A universal signature function* $u : \mathcal{B}_n \times X_n \longrightarrow U$ *is a signature function which has the following property: Given a Boolean function* $f \in \mathcal{B}_n$*, the partition* $\mathcal{A} = \{A_1, A_2, \ldots, A_k\}$ *of its input variables which is induced by* $\mathcal{G}$*-symmetry, and two variables* $x_i, x_j \in X_n$*, then the implication*

$$x_i \in A_l, x_j \in A_h \text{ with } l \neq h \implies u(f, x_i) \neq u(f, x_j)$$

holds.

If two input variables x_i and x_j are not in the same subset of partition $\mathcal{A}$ of a Boolean function $f \in \mathcal{B}_n$, then the universal signature function has to provide different signatures for these two variables. Such a universal signature function has the property to dominate any other signature function $s \in \mathcal{S}_n$:

LEMMA 8.7

$$\forall f \in \mathcal{B}_n \forall x_i, x_j \in X_n : \\ (\exists s \in \mathcal{S}_n : s(f, x_i) \neq s(f, x_j)) \Rightarrow (u(f, x_i) \neq u(f, x_j)) .$$

In other words, if there is any signature function which can distinguish between two input variables of a Boolean function, then the universal signatures of these two variables has to be able to distinguish between them as well.

PROOF: The property of Lemma 8.7 can be easily proved. If two input variables x_i and x_j of a Boolean function $f \in \mathcal{B}_n$ have different signatures for any signature function $s \in \mathcal{S}_n$, then there does not exist an input permutation $\tau \in \mathcal{G}(f)$ with $\tau(i) = j$. Thus, variables x_i and x_j have to be in different subsets of the partition $\mathcal{A}$ of the input variables of f. Because of Definition 8.7, the universal signature of these two variables has to be different as well. ∎

In this sense, we can say that a universal signature function is the strongest signature function which can be constructed.

To close up the investigations let us prove that universal input signature functions really exist.

THEOREM 8.2 *There is a universal signature function* $\mathcal{U} : \mathcal{B}_n \times X_n \longrightarrow U$.

PROOF: At first we construct a candidate for a universal signature function. Then we show, that this candidate is a signature function and that it is universal.

1 Construction:

Let us consider a Boolean function $f \in \mathcal{B}_n$. From Lemma 8.5, it follows that there is a well–defined partition $\mathcal{A}$ of the input variables of function f

which corresponds to the equivalence relation of $\mathcal{R}_{\mathcal{G}(f)}$. Let $\mathcal{A}$ be given by

$$\begin{aligned}\mathcal{A} &= \{A_1, A_2, \ldots, A_k\} \\ &= \{\{x_1^1, \ldots, x_{l_1}^1\}, \{x_1^2, \ldots, x_{l_2}^2\}, \ldots, \{x_1^k, \ldots, x_{l_k}^k\}\}.\end{aligned}$$

Now, we select the *lexicographical* smallest (see Definition 2.10 on page 15) of all functions that can be constructed by permuting the input variables of f:

$$f_{\min} = \min\{g \in \mathcal{B}_n : g = f \circ \tau \text{ with } \tau \in \mathcal{P}er(X_n)\}.$$

This function, $f_{\min}$ is a permutation independent information on f and has a unique partition $\tilde{\mathcal{A}}$ of its input variables as well.

What do we know about the relationship between function f and function $f_{\min}$ and the partition of their input variables? We know that

$$f_{\min} = f \circ \hat{\tau}$$

for an input permutation $\hat{\tau} \in \mathcal{P}er(X_n)$. In other words, these two functions are permutation equivalent. From this fact, it follows that the partition $\tilde{\mathcal{A}}$ has the same structure as partition $\mathcal{A}$:

$$\begin{aligned}\tilde{\mathcal{A}} &= \{\tilde{A}_1, \tilde{A}_2, \ldots, \tilde{A}_k\} \\ &= \{\{y_1^1, \ldots, y_{l_1}^1\}, \{y_1^2, \ldots, y_{l_2}^2\}, \ldots, \{y_1^k, \ldots, y_{l_k}^k\}\}.\end{aligned}$$

Furthermore, there is a one-to-one mapping $\Phi : \mathcal{A} \longrightarrow \tilde{\mathcal{A}}$ which maps the classes of partition $\mathcal{A}$ to the classes of partition $\tilde{\mathcal{A}}$. Φ depends on the input permutation $\hat{\tau}$ which we use to construct $f_{\min}$. Let us have a look at mapping Φ:

Let $\sigma \in \mathcal{G}(f)$ be a permutation which keeps f invariant. Then, the equation

$$\begin{aligned}f_{\min} \circ (\hat{\tau}^{-1} \circ \sigma \circ \hat{\tau}) &= (f \circ \hat{\tau}) \circ (\hat{\tau}^{-1} \circ \sigma \circ \hat{\tau}) \\ &= f \circ (\hat{\tau} \circ \hat{\tau}^{-1}) \circ (\sigma \circ \hat{\tau}) \\ &= f \circ (\sigma \circ \hat{\tau}) \\ &= (f \circ \sigma) \circ \hat{\tau} \\ &= f \circ \hat{\tau} \\ &= f_{\min}\end{aligned}$$

holds, i.e., the permutation $\hat{\tau}^{-1} \circ \sigma \circ \hat{\tau}$ keeps $f_{\min}$ invariant and

$$(\hat{\tau}^{-1} \circ \sigma \circ \hat{\tau}) \in \mathcal{G}(f_{\min}).$$

Furthermore, for any variable x_j, the equation

$$\begin{aligned}(\hat{\tau}^{-1} \circ \sigma \circ \hat{\tau})(\hat{\tau}^{-1}(x_j)) &= (\hat{\tau}^{-1} \circ \sigma \circ \hat{\tau} \circ \hat{\tau}^{-1})(x_j) \\ &= (\hat{\tau}^{-1} \circ \sigma \circ (\hat{\tau} \circ \hat{\tau}^{-1}))(x_j) \\ &= (\hat{\tau}^{-1} \circ \sigma)(x_j) \\ &= \hat{\tau}^{-1}(\sigma(x_j))\end{aligned}$$

holds as well. Note that x_j and $\sigma(x_j)$ are contained in the same class A_i of partition $\mathcal{A}$ because $\sigma \in \mathcal{G}(f)$. The above equation states that $\hat{\tau}^{-1}(x_j)$ and $\hat{\tau}^{-1}(\sigma(x_j))$ are contained in the same class with respect to partition $\tilde{\mathcal{A}}$ as well, because $(\hat{\tau}^{-1} \circ \sigma \circ \hat{\tau}) \in \mathcal{G}(f_{\min})$, and

$$\Phi(A_i) = \hat{\tau}^{-1}(A_i)$$

holds for all $i = 1, \ldots, k$.

Now, let us continue with the construction of a universal signature function. There is a well–defined and ordered partition of the input variables of $f_{\min}$:

$$\mathcal{A}^{f_{\min}} = [A_1^{f_{\min}}, A_2^{f_{\min}}, \ldots, A_k^{f_{\min}}],$$

which is constructed by ordering the classes of partition $\tilde{\mathcal{A}}$ according to Lemma 8.6. With the help of this partition we construct our candidate $\mathcal{U}$ for a universal signature function by setting

$$\mathcal{U}(f, x_i) = j \text{ with } x_{\hat{\tau}^{-1}(i)} \in A_j^{f_{\min}}$$

for any $x_i \in X_n$.

2 Proof of correctness:

We prove that $\mathcal{U}$ is a universal signature function in three steps.

(a) $\mathcal{U}$ is a well–defined mapping from $\mathcal{B}_n \times X_n$ into an ordered set $(U, \leq)$.
Proof: In order to prove that $\mathcal{U}$ is a *well–defined* mapping, we need to consider the case that there are more than one input permutations $\hat{\tau} \in \mathcal{P}er(X_n)$ which construct the lexicographical smallest function

$$f_{\min} = f \circ \hat{\tau}.$$

Suppose, there are two of those input permutations, $\hat{\tau}_1$ and $\hat{\tau}_2$. Then consider any input variable x_i of function f and let $A_r \in \mathcal{A}$ be the class of x_i with respect to f. Both permutations $\hat{\tau}_1^{-1}$ and $\hat{\tau}_2^{-1}$ map A_r onto classes $A_p^{f_{\min}}$ and $A_q^{f_{\min}}$ for some p and q, respectively, i.e.,

$$\hat{\tau}_1^{-1}(A_r) = A_p^{f_{\min}} \text{ and } \hat{\tau}_2^{-1}(A_r) = A_q^{f_{\min}}.$$

Thus, the input permutation $\hat{\tau}_2^{-1} \circ \hat{\tau}_1$ exchanges $A_p^{f_{\min}}$ and $A_q^{f_{\min}}$:

$$\begin{aligned}(\hat{\tau}_2^{-1} \circ \hat{\tau}_1)(A_p^{f_{\min}}) &= \hat{\tau}_2^{-1}(\hat{\tau}_1(A_p^{f_{\min}})) \\ &= \hat{\tau}_2^{-1}(A_r) \\ &= A_q^{f_{\min}}.\end{aligned}$$

Since the equation

$$f \circ \hat{\tau}_1 = f_{\min} = f \circ \hat{\tau}_2$$

holds, the input permutation $\hat{\tau}_2^{-1} \circ \hat{\tau}_1$ keeps the Boolean function $f_{\min}$ invariant as well:

$$\begin{aligned}f_{\min} \circ (\hat{\tau}_2^{-1} \circ \hat{\tau}_1) &= (f \circ \hat{\tau}_2) \circ (\hat{\tau}_2^{-1} \circ \hat{\tau}_1) \\ &= f \circ (\hat{\tau}_2 \circ \hat{\tau}_2^{-1}) \circ \hat{\tau}_1 \\ &= f \circ \hat{\tau}_1 \\ &= f_{\min}.\end{aligned}$$

Thus

$$A_p^{f_{\min}} = A_q^{f_{\min}}$$

holds and $\mathcal{U}$ is well-defined.

(b) $\mathcal{U}$ is a signature function.

Proof: Let us consider the function $f_{\min}$. This function is information about the Boolean function f which is independent of any permutation of the input variables of f since we use *all* $\tau \in \mathcal{P}er(X_n)$ for its construction. Also the partition $\mathcal{A}^{f_{\min}}$ is permutation independent information for f. From this, it follows that the information we obtain for an input variable x_i using $\mathcal{U}$ is independent of any permutation of the inputs.

(c) $\mathcal{U}$ is universal.

Proof: Let us consider the mapping $\Phi : \mathcal{A} \longrightarrow \tilde{\mathcal{A}}$ again, which maps the classes of the partition of the inputs of function f to those of the partition of the inputs of function $f_{\min}$. From the construction of this mapping, it follows that if two input variables x_i and x_j of function f are in different classes of partition $\mathcal{A}$, then $\hat{\tau}^{-1}(x_i)$ and $\hat{\tau}^{-1}(x_j)$ are in different classes of the partition $\tilde{\mathcal{A}}$ of the lexicographical smallest function $f_{\min}$ as well. Remember, the input permutation $\hat{\tau}$ is a permutation of the input variables of function f, which constructs the lexicographical smallest of all those functions $f_{\min} = f \circ \hat{\tau}$.

Now, if we order $\tilde{\mathcal{A}}$ as proposed in Lemma 8.6, then we obtain the well–defined and ordered partition $\mathcal{A}^{f_{\min}}$ for the function $f_{\min}$, and it follows for any Boolean function $f \in \mathcal{B}_n$ and for all input variables $x_i, x_j \in X_n$:

$$\begin{aligned} & x_i \in A_l \text{ and } x_j \in A_h \text{ with } l \neq h \\ \Leftrightarrow \quad & x_{\hat{\tau}^{-1}(i)} \in A_p^{f_{\min}} \text{ and } x_{\hat{\tau}^{-1}(j)} \in A_q^{f_{\min}} \text{ with } p \neq q. \end{aligned}$$

which is equivalent to

$$\begin{aligned} & x_i \in A_l \text{ and } x_j \in A_h \text{ with } l \neq h \\ \Leftrightarrow \quad & \mathcal{U}(f, x_i) \neq \mathcal{U}(f, x_j). \end{aligned}$$

■

The signature function $\mathcal{U}$ demonstrates the property of a universal signature function to dominate all other signature functions very well. The partition $\mathcal{A}$, which is the basic component for the construction of $\mathcal{U}$, is nothing else but the partition of the input variables which we try to construct with the help of our practical signatures. Now, if the universal signature for an input variable is the index of such a class of the partition, then one fact is obvious. If there is a practical signature which is different for two variables, then these two variables will be in two different classes of the partition. Thus, the universal signatures of these two variables will be different as well

Why does it help to know that there is a universal signature function? It cannot be used in practice since its construction would be computationally too intensive. However, it is useful to have such a universal signature function for theoretical investigations. In [85], a method for analytically comparing the effectiveness of signature functions is introduced. Given a signature function $s : \mathcal{B}_n \times X_n \longrightarrow U$, a measure $\mu(s)$ of the effectiveness of s is the cardinality of the co–domain of s which is $|U|$. However, as we can see here, a small co–domain does not automatically imply that the signature function is not efficient. In the case of the universal signature function, the inequation $\mu(\mathcal{U}) \leq n$ holds. However, as we have proven, the universal signature function is the most powerful signature function at all.

Let us discuss now signatures in relationship to $\mathcal{G}$–symmetry. Because of Definition 8.7, if any two input variables of a Boolean function have the same signature with respect to the universal signature function, then there is a non-trivial permutation $\tau \in \mathcal{P}er(X_n) \setminus \{\mathbf{1}\}$ of the input variables of this function, such that $f = f \circ \tau$. Since this permutation τ of the input variables keeps the function f invariant, the function f is $\mathcal{G}$–symmetric with respect to a group $\mathcal{G}$ of input permutations with $\mathcal{G} \neq \{\mathbf{1}\}$ and Theorem 8.3 holds.

THEOREM 8.3 *Let $f \in \mathcal{B}_n$ and let $\mathcal{A} = \{A_1, A_2, \ldots, A_k\}$ be the partition of the input variables of f induced by $\mathcal{G}$-symmetry. Then*

$$\forall x_i, x_j \in X_n : (\mathcal{U}(f, x_i) = \mathcal{U}(f, x_j)) \\ \Longrightarrow (\exists l \in \{1, \ldots, k\} : x_i, x_j \in A_l)\,.$$

Together with Theorem 8.1, the property of a universal signature ensures that we do not need to think about other problems with signatures identifying input variables for permutation independent function equivalence. Assuming the availability of efficient working signature functions, it demonstrates that G–symmetry is indeed the only handicap for signatures to be able to uniquely identify an input variable of a Boolean function independent of permutation. Thus, we now need to focus on $\mathcal{G}$–symmetries.

5. Special kinds of $\mathcal{G}$–symmetries

Specifically, we now discuss some special kinds of $\mathcal{G}$–symmetry which often appear in practice. Of course, this cannot be a complete enumeration of possible $\mathcal{G}$–symmetries. These cases have been discovered in the quest to understand why signatures were proving to be inadequate for permutation independent Boolean comparison in some cases.

5.1 Partial symmetries

At first, let us consider the most common case. As already defined in Definition 8.5 on page 182, a Boolean function $f \in \mathcal{B}_n$ is *symmetric* in a subset of input variables $\hat{X} \subseteq X_n$ if f is invariant under all permutations of the input variables in $\hat{X}$. We say, that f is *symmetric* (or *partial symmetric*, in order to be able to differ between this special kind of $\mathcal{G}$-symmetry and other kinds of $\mathcal{G}$-symmetry) in $\hat{X}$. Furthermore, the set $\hat{X}$ is a *maximal symmetry group* (a maximal set of symmetric variables) of f if f is symmetric in $\hat{X}$, and there is no variable $x_i \notin \hat{X}$ such that f is symmetric in $\hat{X} \cup \{x_i\}$ as well. This is the simplest kind of $\mathcal{G}$–symmetry, and it is well–known. In the partition $\mathcal{A}$ of the inputs of a Boolean function f which is partial symmetric in $\hat{X}$, all input variables of subset $\hat{X}$ are in one subset.

Partial symmetries are easy to detect and to handle.

THEOREM 8.4 *A Boolean function f is partially symmetric in the variables x_i and x_j if and only if $f_{[x_i,x_j]=(1,0)} = f_{[x_i,x_j]=(0,1)}$.*

With the help of this theorem we are able to test partial symmetry, since the correspondent relation is an equivalence relation [110]. Furthermore, different fast methods to improve this basic symmetry test have been developed. We use the methods introduced in [110], while one of the methods introduced there is to use simple signatures. For more details, please see [110, 132].

After the detection of partial symmetries we are done with that special kind of $\mathcal{G}$–symmetry, since each correspondence between symmetric variables of two Boolean functions is fine for the purpose to test permutation equivalence. Thus, one of the first steps of the identification process is to determine all maximal groups of pairwise symmetric variables. The advantage is that further signature computations can be restricted to one representative of each maximal symmetry group. We suggest to determine the maximal symmetry groups *after* applying the cofactor satisfy count signature function. Establishing a first variable partition by using this signature function is computationally not as intensive as applying the symmetry check to each pair of variables. Furthermore, we then have to apply the symmetry check to variables which are element of the same aliasing group only. By this, we combine methods of symmetry detection with the advantages of applying different signature functions.

5.2 Hierarchical symmetries

Investigations on our benchmark set have shown that for several examples the reason for the existence of aliasing groups after computation of all signatures functions is the following kind of symmetry.

DEFINITION 8.8 (HIERARCHICAL SYMMETRY) *Let $f \in \mathcal{B}_n$ be a Boolean function with the input variables $X_n = [x_1, x_2, \ldots, x_n]$ and let $Y_1, Y_2 \subset X_n$ be two subsets of X_n. Y_1 and Y_2 are hierarchical symmetric (h–symmetric) if and only if*

- $|Y_1| = |Y_2| > 1$
- *Y_1 and Y_2 are maximal symmetry groups of f.*
- *f is $H(Y_1, Y_2)$–symmetric, where $H(Y_1, Y_2)$ is the subgroup of the permutation group $\mathcal{P}er(X_n)$ generated by the set*

  $$H(Y_1, Y_2) = \{\tau \in \mathcal{P}er(X_n) |\ \tau(Y_1) = Y_2 \text{ and } \tau(Y_2) = Y_1\}$$

  of input permutations, i.e., f keeps invariant under any exchanging of the variables of Y_1 with those of Y_2.

A group of subsets $\{Y_1, Y_2, \ldots, Y_k\}$ of X_n is h-symmetric if and only if for all $i, j \in \mathbb{N}_k$ set Y_i is h–symmetric to set Y_j.

For illustration, consider the Boolean function

$$f = (x_1 + x_2)' + (x_3 + x_4)' + (x_5 + x_6)'.$$

Here, $\{x_1, x_2\}$, $\{x_3, x_4\}$, and $\{x_5, x_6\}$ are pairs of symmetric variables, but there is no partial symmetry between two variables of different pairs. However, it is easy to see, that exchanging any two of these three pairs keeps the function f invariant. This simple example illustrates h–symmetry.

LEMMA 8.8 *The h–symmetry relation is an equivalence relation on the partition of the set of input variables of a Boolean function f in its maximal symmetry groups.*

PROOF: The symmetry and reflexivity is obviously to see. For transitivity we have to show the following. Let Y_1, Y_2, and Y_3 be three disjoint sets of symmetric variables. If Y_1 is h–symmetric to Y_2 and Y_2 is h–symmetric to Y_3, then Y_1 and Y_3 are h–symmetric as well. Therefore, let us go through all points of the definition of h–symmetry:

- $|Y_1| = |Y_2| = |Y_3| > 1$
- true by assumption
- We know, that exchanging Y_1 with Y_2 as well as exchanging Y_2 and Y_3 does not change the function f. So, let us do the following. First, exchange the variables of Y_1 with those of Y_2. After that, exchange the variables of Y_2 with those of Y_3. The resulting function is equal to f due to h–symmetry of Y_1 and Y_2 and h–symmetry of Y_2 and Y_3. What we have done is to exchange the variables of Y_1 with those of Y_3 using the variables of Y_2. This implies that exchanging Y_1 with Y_3 does not change function f.

∎

The definition of h–symmetry indicates that this is a special kind of $\mathcal{G}$–symmetry. Let us examine the partition $\mathcal{A}$ of the input variables constructed by h–symmetry.

THEOREM 8.5 *If two subsets Y_1 and Y_2 of input variables are h–symmetric then all input variables of Y_1 and Y_2 are in one element A_i of partition $\mathcal{A}$.*

PROOF: This follows directly from the definition of partition $\mathcal{A}$. ∎

Thus, from Theorem 8.1, we know that all of these variables have to have the same signature, i.e., they form an aliasing group. In other words, there is no way to distinguish between them via signatures. However, partly there is a solution for this problem in practice. It is based upon our handling of symmetric variables. To understand this, let us consider the algorithm to identify the input variables by signatures. Here, we first determine all maximal groups of pairwise symmetric variables (see previous section). In this way, pairwise symmetric variables are kept together in aliasing groups. Now, let us consider two h–symmetric subsets Y_1, Y_2 of input variables of function f which form an aliasing group $Y_1 \cup Y_2$. A correspondence between these variables and the variables of an aliasing group Z of any other function g is possible if the variables of aliasing group Z have the same signature than the variables of $Y_1 \cup Y_2$ and aliasing group Z has the same structure than aliasing group $Y_1 \cup Y_2$. That is, there are maximal symmetry groups Z_1 and Z_2 with $Z = Z_1 \cup Z_2$, $|Z_1| = |Y_1|$, and $|Z_2| = |Y_2|$. Then there are two possible correspondences between these groups:

$$(Y_1 \leftrightarrow Z_1, Y_2 \leftrightarrow Z_2)$$

as well as

$$(Y_1 \leftrightarrow Z_2, Y_2 \leftrightarrow Z_1).$$

Because of h–symmetry both correspondences are acceptable for our purpose. In other words, the remaining task in terms of h–symmetry is to *detect* this kind of symmetry. This is sufficient to decide that no further work needs to be done with these aliasing groups in order to solve the permutation problem.

So let us try and see what we have to do. At first, we answer the following question. Let f be a Boolean function with n input variables, $X_n = [x_1, x_2, \ldots, x_n]$, and Y_1 and Y_2 be two disjoint subsets of the set of inputs, i.e., $Y_1, Y_2 \subset X_n$. When is it possible to exchange Y_1 and Y_2 in the function f without changing f itself? An obvious necessary condition is that the number of variables in Y_1 and Y_2 has to be the same. Otherwise one subset cannot be completely exchanged with the other. However, this is not sufficient. Supposing that $|Y_1| = |Y_2| = k$, let us see what sufficient condition exists for the exchangeability of Y_1 and Y_2.

Let $f_{[Y_1,Y_2]=(\alpha^{(1)},\alpha^{(2)})}$ be the cofactor of f where the variables of Y_1 are set to $\alpha^{(1)} \in \{0,1\}^k$ and the variables of Y_2 are set to $\alpha^{(2)} \in \{0,1\}^k$. A sufficient condition is the following:

THEOREM 8.6 *Exchanging two different, ordered subsets of variables,* $Y_1 = [x_1^{(1)}, \ldots, x_k^{(1)}]$ *and* $Y_2 = [x_1^{(2)}, \ldots, x_k^{(2)}]$*, i.e., exchanging* $x_i^{(1)}$ *with* $x_i^{(2)}$ *for all* $i \in \mathbb{N}_k$*, does not change the function* f *if and only if for all assignments* $\alpha^{(1)}, \alpha^{(2)} \in \{0,1\}^k$ *the equation*

$$f_{[Y_1,Y_2]=(\alpha^{(1)},\alpha^{(2)})} = f_{[Y_1,Y_2]=(\alpha^{(2)},\alpha^{(1)})}$$

holds.

Note that a well–known special case of this property is $k = 1$, i.e., the question whether two variables x_i and x_j are exchangeable in a Boolean function f without changing f. We know that this kind of symmetry can be shown using

Theorem 8.4. In other words, Theorem 8.6 is a generalization of the well–known symmetry test for symmetric variables to a test for the exchangeability of groups of variables.

However, since both, Y_1 and Y_2 are symmetry groups, it is much easier to test whether Y_1 and Y_2 can be exchanged without changing f than for the general case stated in Theorem 8.6. We know that f keeps invariant under all permutations of the variables in Y_i ($i \in \{1,2\}$). Furthermore, each vector $\alpha \in \{0,1\}^k$ with exactly l one's in it is a permutation of any other vector with l one's in it. Thus, f is partial symmetric in Y_i if and only if f depends on the number of one's in the input assignment to Y_i only [146]. We will call the number of one's in an input assignment to a set Y_i of variables the *weight* of that input assignment with respect to Y_i. Thus, we can directly conclude the following two facts about the cofactors of f with respect to a set of symmetric variables.

LEMMA 8.9 *Let V_l be the set of all assignments to the set Y_i of symmetric variables with weight l. Let $\alpha^{(1)}, \alpha^{(2)} \in V_l$. Then $f_{Y_i=\alpha^{(1)}} = f_{Y_i=\alpha^{(2)}}$ holds.*

This gives us the number of different cofactors with respect to a set Y_i of symmetric variables that can be constructed.

LEMMA 8.10 *Consider the set of the cofactors of f that can be constructed with respect to the set Y_i of symmetric variables. The cardinality of this set is bounded by $|Y_i| + 1$.*

With Lemmas 8.9 and 8.10 the test of Theorem 8.6 which checks whether Y_1 and Y_2 are exchangeable without changing function f can efficiently be performed. We have to consider $k+1$ assignments to Y_1 as well as to Y_2, one of every possible weight. In other words, we do not have to test all combinations of the 2^k *assignments* but only all combinations of the $k+1$ possible *weights*.

THEOREM 8.7 *Exchanging two different subsets Y_1 and Y_2 of symmetric variables does not change function f if and only if for all weights $l_1, l_2 \in \{0,\ldots,k\}$ with $l_1 \neq l_2$ the equation*

$$f_{[Y_1,Y_2]=(\zeta_{l_1}^{(k)},\zeta_{l_2}^{(k)})} = f_{[Y_1,Y_2]=(\zeta_{l_2}^{(k)},\zeta_{l_1}^{(k)})}$$

with $\zeta_w^{(k)} = (\underbrace{1,\ldots,1}_{w},\underbrace{0,\ldots,0}_{k-w})$ *holds.*

Note, that we have to know that Y_1 and Y_2 are sets of symmetric variables before using Theorem 8.7 for our purpose.

What about the weight combinations $0/0, 1/1, \ldots, k/k$? Obviously, we do not have to test these combinations, since an exchange of the two variable groups with these assignments does not change the function value of f.

Suppose the symmetry of the variables of our two groups Y_1 and Y_2 is tested before, we obtain the moderate number of $(k^2 + k)/2$ tests necessary to check the h–symmetry of Y_1 and Y_2, namely k choices for the assignment to Y_2 in the first step, $k - 1$ in the second, and so on. Dealing with ROBDDs we know that these tests need constant time on the given cofactors only. The cofactor construction itself needs time linear in the number of ROBDD nodes of f. So, in all, we need time $O(k^2 \cdot |F|)$ where $|F|$ denotes the number of nodes in the ROBDD F of f and k the size of the two variable groups we want to test.

Now we can start with a description of the complete algorithm to determine the h-symmetry groups of a Boolean function $f \in \mathcal{B}_n$. First, let us illustrate the algorithm by example. Again, let us consider the Boolean function

$$f = (x_1 + x_2)' + (x_3 + x_4)' + (x_5 + x_6)'$$

which is h-symmetric in $\{\{x_1, x_2\}, \{x_3, x_4\}, \{x_5, x_6\}\}$. This can be proved as follows:

1. Consider the groups of symmetric variables. In the case of this example there are three groups with size 2, namely $Y_1 = \{x_1, x_2\}$, $Y_2 = \{x_3, x_4\}$, and $Y_3 = \{x_5, x_6\}$.

2. To prove the h–symmetry of these groups it is enough to check Y_1 with Y_2 and Y_2 with Y_3, because of Lemma 8.8. Three cofactor tests are necessary in both cases, namely for the weight combinations 0/1, 0/2, and 1/2. For Y_1 and Y_2 these are

$$\begin{aligned}
f_{[x_1,x_2,x_3,x_4]=(0,0,1,0)} &= f_{[x_1,x_2,x_3,x_4]=(1,0,0,0)} = \mathbf{1}, \\
f_{[x_1,x_2,x_3,x_4]=(0,0,1,1)} &= f_{[x_1,x_2,x_3,x_4]=(1,1,0,0)} = \mathbf{1}, \text{ and} \\
f_{[x_1,x_2,x_3,x_4]=(1,0,1,1)} &= f_{[x_1,x_2,x_3,x_4]=(1,1,1,0)} = (x_5 + x_6)'.
\end{aligned}$$

Analogously, three cofactor tests have to be done to check the exchangeability of Y_2 and Y_3.

Now, let us come to our algorithm. Given a Boolean function $f \in \mathcal{B}_n$ we can determine the h–symmetries of the variables of f in two steps.

- In the **first step** the list of all candidates for h–symmetry is computed. A *candidate for h–symmetry* is a set of same–sized partial symmetry groups of variables of f, except those of size one. Note, that we construct maximal candidates only. So, there is no h–symmetry possible between different candidates. In the above example, there is exactly one candidate, namely $\{\{x_1, x_2\}, \{x_3, x_4\}, \{x_5, x_6\}\}$.

- In the **second step** we use Theorem 8.7 and Lemma 8.8 for each candidate to establish the h–symmetry groups.

Step 2 of this approach has been described already on the above example. The generalization is straightforward. Note that a candidate for h–symmetry could include sets of input variables that belong to different h–symmetries. If this case has to be taken into account, then we need to test the exchangeability of *each* pair of variable sets in the candidate.

Let us discuss Step 1. What is the best way to determine the list of candidates? One way is the following. First, compute all groups of partial symmetries of X_n with respect to function f, then sort these groups by their size. A candidate is the set of all symmetry groups with the same size, except the one including all groups of size one. However, this kind of selection of candidates is not the best – any cofactor test (and so any cofactor construction) that can be avoided is a bonus point with respect to CPU time and storage requirements. Nevertheless, we already *know* a better way for this. Theorem 8.1 tells us that the variables of any two partial symmetry groups have to have the same signature to be a candidate for h–symmetry. That is why the signature computation is an efficient pre–processing for the selection of candidates for h–symmetry. Only aliasing groups which consist of groups of symmetric variables are candidates. In this way we are likely to obtain a smaller set of candidates, such that we can avoid cofactor constructions. Thus to create the list of candidates for h–symmetry, we use all the signatures first. If there are aliasing groups, then we select those groups that consist of symmetric variable groups as candidates for h–symmetry.

5.3 Group symmetries

Now we change our focus to the following kind of symmetry.

DEFINITION 8.9 (GROUP SYMMETRY) *Let $f \in \mathcal{B}_n$ be a Boolean function with n input variables $X_n = [x_1, x_2, \ldots, x_n]$. Let $Y_1, Y_2, \ldots, Y_k \subset X_n$ be non-empty and pairwise disjoint subsets of X_n with $k > 1$. The k groups of input variables are group symmetric (g–symmetric) if and only if*

1 $|Y_i| > 1$ for all $i \in \mathbb{N}_k$.

2 There are (non-trivial) permutations

$$\tau_1, \tau_2, \ldots, \tau_k$$

of the variables in

$$Y_1, Y_2, \ldots, Y_k,$$

respectively, such that applying the input permutations τ_i to Y_i simultaneously for all $i \in \mathbb{N}_k$ *does not change the function* f.

Note, that manipulating just one or a couple of variable subsets Y_i may change the function f. The following example will help to clarify the definition. Consider the Boolean function

$$f = (a_0 \cdot x_1{}' \cdot x_2{}') + (b_0 \cdot x_3{}' \cdot x_4{}').$$

Here, $\{x_1, x_2\}$ and $\{x_3, x_4\}$ are pairs of symmetric variables, but there is no h-symmetry between them because of the existence of the input variables a_0 and b_0. However, exchanging $\{x_1, x_2\}$ and $\{x_3, x_4\}$ *and* a_0 and b_0 keeps the function invariant. So, there is what we call g–symmetry between the two subsets of input variables, $\{x_1, x_2, x_3, x_4\}$ and $\{a_0, b_0\}$, and

$$\mathcal{A} = \{\{x_1, x_2, x_3, x_4\}, \{a_0, b_0\}\}$$

holds.

Again, we have a case where signatures will be unable to distinguish between the variables and thus results in the formation of aliasing groups. Unfortunately, practical experiences show that this kind of $\mathcal{G}$–symmetry appears relatively often. One well–known example is an n–bit multiplier, which computes

$$\phi_2^{-1}\left(\phi_2(x_n, x_{n-1}, \ldots, x_{\frac{n}{2}+1}) \cdot \phi_2(x_{\frac{n}{2}}, \ldots, x_2, x_1)\right),$$

where $\phi_2 : \{0,1\}^\star \to \mathbb{Z}$ is the two's complement encoding – we assume that n is an even natural number. Exchanging $x_{\frac{n}{2}+i}$ and x_i for each $i = 1, \ldots, \frac{n}{2}$ simultaneously keeps the function invariant. The partition $\mathcal{A}$ of these variables is

$$\{\{x_i, x_{\frac{n}{2}+i}\} |\ i = 1, \ldots, \frac{n}{2}\}.$$

So there are at least $2^{\frac{n}{2}}$ possible variable correspondences for an n-bit multiplier after signature computation. Taking the g–symmetry of the n–bit multiplier function into account, this number could decrease by one half, i.e., instead of $2^{\frac{n}{2}}$ possible variable correspondences, only $2^{\frac{n}{2}-1}$ correspondences have to be checked, which is still too large in general.

Moreover, it seems to be very complicated to detect a g–symmetry in general. To be able to handle general g–symmetries, we need a way to detect g–symmetric groups of input variables *and* a way to select those permutations of these variables that keep the g–symmetric function invariant. Considering the very general property of g–symmetry, this looks like a complicated task.

In looking for possible ways to handle g–symmetry we made the following interesting observation. The subsets of input variables that result in a g–symmetry are often connected with each other in the following sense: if we have identified the variables of one of these subsets, then it is possible to identify the variables of the other subsets as well. With this knowledge we can develop heuristics to distinguish between variables of g–symmetric subsets. It works as follows. Given a Boolean function $f \in \mathcal{B}_n$ let us consider the following situation. There is a partial, permutation independent order of the n input variables of f constructed by using the signatures presented above. Furthermore, the groups of symmetric input variables (see Section 5.1) are identified as well as the aliasing groups with h-symmetric groups of input variables (see Section 5.2). Nevertheless, there are still unidentified groups of aliasing variables. For the sake of simplicity, say these groups are $A_1, A_2, \ldots, A_k$. In this situation, we assume that all k groups of aliasing variables are connected by one g–symmetry and use the observation that the groups of input variables that result in a g–symmetry are often connected with each other in the special sense just mentioned.

Now, we just *assume* that the input variables of one of these aliasing groups, say A_1, are uniquely identified. Under this assumption, we apply a couple of function signatures to each variable of the other $k-1$ aliasing groups, i.e., we construct function signatures that depend not only on those input variables that can be uniquely identified but also on those of aliasing group A_1. The basic idea of this heuristic is that we may be able to find a unique function signature for each of the input variables in A_2 to A_k in the case of g–symmetry, because of the connection among all k aliasing groups of such a g-symmetry. If this is the case, we can uniquely identify each input variable in the aliasing groups A_2 to A_k. Practical experiences have shown that often this is indeed the case [105].

Let us consider an example. Consider the Boolean function

$$\begin{aligned} & f(a_1, a_2, a_3, x_1, x_2, x_3) \\ & \quad = \quad (a_1 \cdot (x_1 + x_2)') + (a_2 \cdot (x_2 + x_3)') + (a_3 \cdot (x_3 + x_1)'). \end{aligned}$$

This function f is g–symmetric with respect to the variable groups $\{a_1, a_2, a_3\}$ and $\{x_1, x_2, x_3\}$. Let us see how the heuristics works. We pick up one of these two aliasing groups, say $\{a_1, a_2, a_3\}$, and assume that we can uniquely identify these three variables. We assume that $a_1 < a_2 < a_3$ holds. Under this assumption we can apply a function signature to the other three variables which may depend on a_1, a_2, and a_3:

$$\begin{aligned} f_{[x_1,x_2,x_3]=(1,0,0)} &= a_2, \\ f_{[x_1,x_2,x_3]=(0,1,0)} &= a_3, \\ f_{[x_1,x_2,x_3]=(0,0,1)} &= a_1. \end{aligned}$$

And indeed, the variables x_1, x_2, and x_3 can be uniquely identified with these signatures. Furthermore, we can use the order relation for function signatures to obtain a unique order of the variables x_1, x_2, and x_3. Applying the lexicographically order relation for Boolean functions (see Definition 2.10 on page 15) the unique order of the three variables is (x_3, x_1, x_2).

This unique description is not permutation independent. It depends on the permutation of the input variables in A_1, which was in the case of the above example (a_1, a_2, a_3). So we need a way to make the information permutation independent. For this, we consider the permutation τ of the input variables of f which results into the current unique order of these variables given by the signatures (assuming that there is a unique ordering). In the above example, this input permutation $\tau \in \mathcal{P}er(\{a_1, a_2, a_3, x_1, x_2, x_3\})$ is defined by

$$\tau(y) = \begin{cases} a_i, & \text{if } y = a_i \text{ for some } i \in \{1, 2, 3\} \\ x_3, & \text{if } y = x_1 \\ x_1, & \text{if } y = x_2 \\ x_2, & \text{if } y = x_3. \end{cases}$$

With the help of τ, we construct the Boolean function $\hat{f} = f \circ \tau$. This function is stored in a set $\hat{F}$. Now, the same procedure is carried out for each possible order of the input variables in A_1.

Table 8.1 lists all the orders of $A_1 = \{a_1, a_2, a_3\}$ in the left column. Each of these orders implies an order for the variables $\{x_1, x_2, x_3\}$ which are listed in the right column of the table.

Table 8.1. Group Symmetry: Computation of $\hat{F}$

Order of $\{a_1, a_2, a_3\}$	Induced order of $\{x_1, x_2, x_3\}$
(a_1, a_2, a_3)	(x_3, x_1, x_2)
(a_1, a_3, a_2)	(x_3, x_2, x_1)
(a_2, a_1, a_3)	(x_1, x_3, x_2)
(a_2, a_3, a_1)	(x_1, x_2, x_3)
(a_3, a_1, a_2)	(x_2, x_3, x_1)
(a_3, a_2, a_1)	(x_2, x_1, x_3)

After computation of $\hat{F}$, the lexicographical smallest of all Boolean functions in $\hat{F}$ is selected. The variable order which belongs to this function and the function itself are the permutation independent information for function f.

Note, that we can also use any of the other sets of aliasing variables, $A_2, \ldots, A_k$. We suggest to use one of those sets with minimal size, since this reduces the number of steps necessary to construct set $\hat{F}$.

5.4 The variety of $\mathcal{G}$–symmetry

We introduced three kinds of $\mathcal{G}$–symmetry that cover a wide range of symmetries appearing in practice. For these symmetries there are efficiently working solution paradigms for the permutation equivalence problem [105]. Nevertheless, it seems to be not possible to find a general solution paradigm which allows us to handle $\mathcal{G}$–symmetry efficiently because of the generality of this function property. Each kind of $\mathcal{G}$–symmetry seems to demand a special procedure to handle it. Although we feel that the $\mathcal{G}$–symmetries introduced here are those mostly appearing in practice, other symmetries may appear as well.

Let us present one of these $\mathcal{G}$–symmetries. Consider the Boolean function

$$f = (x_1 \cdot x_2) + (x_2 \cdot x_3) + (x_3 \cdot x_4) + (x_4 \cdot x_5) + (x_5 \cdot x_1).$$

It is obvious that rotating the five variables in function f does not change this function. So it is a $\mathcal{G}$–symmetry as well. All five input variables of the Boolean function f will form one aliasing group that cannot be refined by using signatures. Furthermore, there are no partial, no hierarchical, as well as no group symmetries between these input variables. We call this kind of symmetry *rotational symmetry*.

DEFINITION 8.10 (ROTATIONAL SYMMETRY) *Let $f \in \mathcal{B}_n$ be a Boolean function. Let*

$$Y = [y_1, y_2, \ldots, y_k] \subseteq X_n$$

be a subset of the input variables of f. Let $\tau \in \mathcal{P}er(X_n)$ be the permutation on X_n which is defined by

$$(\forall x_i \in X_n)\ \tau(x_i) = \begin{cases} y_{(j \bmod k)+1}, & \text{if } x_i = y_j \text{ for some } j \in \mathbb{N}_k \\ x_i, & \text{if } x_i \notin Y \end{cases}$$

f is rotational symmetric (r–symmetric) if and only if

- *$k \geq 3$,*
- *f is not partial symmetric in Y, and*
- *f does not change on applying the input permutation τ to the variables of Y.*

Note that also the input permutation τ^{-1} keeps f invariant as well as applying τ more than once, since the set $\mathcal{G} \subset \mathcal{P}er(X_n)$ which is constructed by an r–symmetry is a group. Similar to group symmetry, r–symmetry is neither easy to detect nor easy to handle. Furthermore, the heuristic applied for the group symmetries, (i.e., assume one of the input variables of Y has been uniquely identified by a signature, then try to use function signatures to distinguish between the others as well) does not work very well for rotational symmetry, as experiments have shown. Rotational symmetry illustrates the difficulties that may appear in general with $\mathcal{G}$–symmetries.

However, partial symmetries, hierarchical symmetries, and group symmetries seem to be the most common $\mathcal{G}$–symmetries appearing in practice. So, the algorithms presented for handling these symmetries have direct practical impact to the complete solution of the permutation problem. For some insights into practical experiences with using these techniques we refer to [105].

Notes

1 As already said in connection with Definition 8.1 on page 171, a permutation $\tilde{\tau}$ of $\mathbb{N}_n$ induces a permutation $\overline{\tau} \in \mathcal{P}er(X_n)$ defined by

$$\forall x_i \in X_n : \overline{\tau}(x_i) = x_{\tilde{\tau}(i)},$$

and vice versa. It also induces a permutation $\tau \in \mathcal{P}er(\{0,1\}^n)$ defined by

$$\tau(x_1, \ldots, x_n) = (x_{\tilde{\tau}(1)}, \ldots, x_{\tilde{\tau}(n)}).$$

For the sake of simplicity, we use $\tilde{\tau}$, τ, and $\overline{\tau}$ interchangeably and use the notation τ for each of the three functions.

III

EQUIVALENCE CHECKING OF SEQUENTIAL CIRCUITS

The current output of digital circuits may depend not only on the current assignment of the primary inputs but on previous values also. Sequential circuits allow to store information of the history of previous values in internal storage, i.e., flipflops, registers, and random access memories. The current values of these storage elements constitute the state of the sequential circuit.

In theory, any digital system can be modelled as a finite-state machine (FSM). There is a well-developed theory for analyzing FSMs, including checking their equivalence [75]. We are going to give a detailed introduction to this formal basics in Chapter 9. Section 3 of Chapter 9 is going to dwell on state space traversal which is the basis of most of the general equivalence checking methods for sequential circuits.

In most cases, equivalence checking is applied to sequential circuits to show that certain circuit modifications have not altered the functional behavior of the circuit. Often, the state encoding of both circuits remains the same. In this special case it is not necessary to perform a costly state space traversal. It is sufficient to compare the Boolean output function and a state transition function which can be considered as combinational circuits when the correspondence of the latches of the two sequential circuits to be checked can be established. As already explained in the motivation of permutation independent Boolean comparison (Chapter 8), there are several cases in with the correspondence may get lost. Here the verifier has to be able to establish the latch correspondence based on the individual functions to avoid time consuming state space traversal [118]. Chapter 10 is going to deal with this latch correspondence problem.

Chapter 9

FORMAL BASICS

In Section 1 basic notations with regard to the theory of sequential circuits are reviewed. In particular, we concentrate on the definition of finite state machines and their representations. Section 2 gives a formal definition of the equivalence of finite state machines and reviews the traditional basic methods to compute state equivalences. Algorithms for traversing the state space of a finite state machine are presented in Section 3. They are the basis of most of the equivalence checking procedures for sequential circuits known in literature. The chapter closes with error trace generation.

1. Finite State Machines and their representations

Finite state machines are critical for realizing the ´´decision-making´´ logic in digital logics. Since in hardware implementations, nondeterminism should not appear, we only consider deterministic finite automata, in the following.

1.1 Basic definitions

There are two basic ways to organize clocked sequential circuits, Moore automata and Mealy automata.

DEFINITION 9.1 (MOORE AUTOMATON)
A Moore automaton is a tuple $M = (S, I, O, \delta, \lambda, r)$. It consists of

- *a finite set $S \subseteq \{0,1\}^q$ of states for some constant $q \in \mathbb{N}$,*
- *a finite set $I \subseteq \{0,1\}^n$ of input symbols for some constant $n \in \mathbb{N}$,*
- *a finite set $O \subseteq \{0,1\}^m$ of output symbols for some constant $m \in \mathbb{N}$,*

- *a transition function* $\delta : S \times I \rightarrow S$ *which takes as arguments a state* s *and an input symbol* w *and returns the state that the automaton enters after processing input symbol* w *in state* s.

- *an output function* $\lambda : S \rightarrow O$ *which specifies the output of the automaton depending on the present state, and*

- *a start state* $r \in S$.

Figure 9.1 illustrates the definition of a Moore automaton. Here, we consider

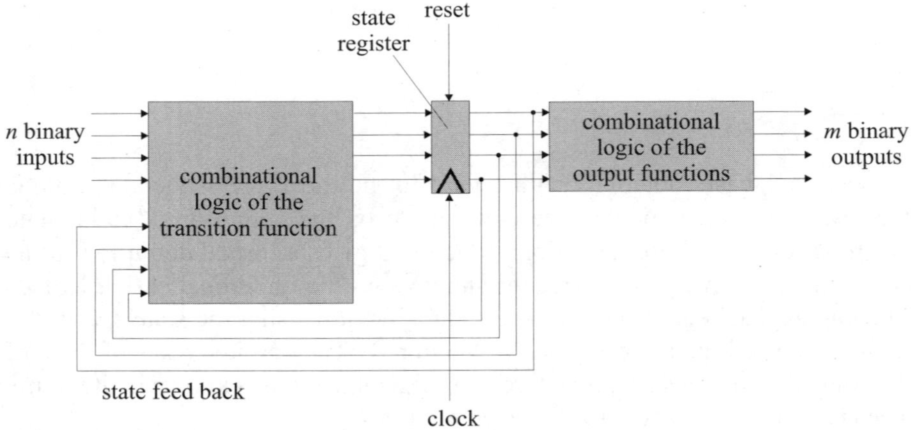

Figure 9.1. Moore machine [70]

sequential circuits with reset lines, i.e., circuits which are in a definite initial state r after power-on. Outputs are computed by a combinational logic circuit whose only inputs are the state outputs, which we denote as *present state variables* in the following. Thus the outputs change synchronously with the state transition and the clock edge. Both facts are desirable properties of real hardware controllers because fully synchronous finite state machines with reset lines are much easier to implement and debug than asynchronous automata without reset lines. Furthermore, since we focus on given hardware implementations of such automata, states, input symbols, and output symbols are considered to be binary encoded.

Moore automata are special Mealy automata which are a more general concept, from a practical point of view. Often, a Mealy automaton can generate the same output sequence in fewer states than a Moore automaton.

DEFINITION 9.2 (MEALY AUTOMATON) *A Mealy automaton is a tuple* $M = (S, I, O, \delta, \lambda, r)$. *S, I, O,* δ, *and* r *are defined as in Definition 9.1. Output function* λ *is a function that specifies the output of the automaton depending on the present state and the present value of the input, i.e.,* $\lambda : S \times I \to O$.

The outputs of Mealy automata depend on the present state and the present value of the primary inputs. Figure 9.2 illustrates this inherent asynchronous behavior. The outputs can change immediately after a change at the inputs, independent of

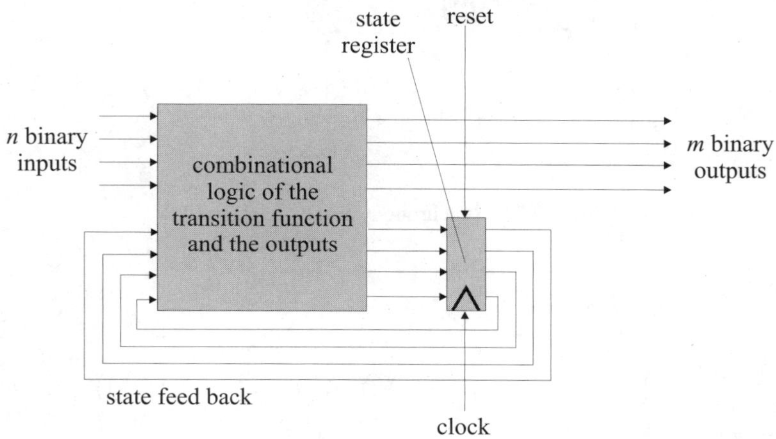

Figure 9.2. Mealy machine [70]

the clock. Such glitches are undesirable in real hardware controllers. This leads to an alternative synchronous design style for Mealy automata by synchronizing the Mealy automata outputs with output flipflops as shown in Figure 9.3. In this book, we concentrate on such synchronous Mealy automata, which we call Finite State Machines (FSM). Note that there should be no glitches at the outputs of synchronous Mealy automata.

Since we are interested in what happens when we start in a state s and follow a finite sequence $w \in I^+$ of input symbols, we define the *extended transition function*

$$\delta^+ : S \times I^+ \to S$$

and extended output function

$$\lambda^+ : S \times I^+ \to O^+$$

which are defined for all $s \in S$ and $w \in I^+$ as

$$\delta^+(s, w) = \begin{cases} \delta(s, w) & \text{, if } |w| = 1 \\ \delta(\delta^+(s, w_{1\ldots|w|-1}), w_{|w|}) & \text{, if } |w| \geq 2 \end{cases}$$

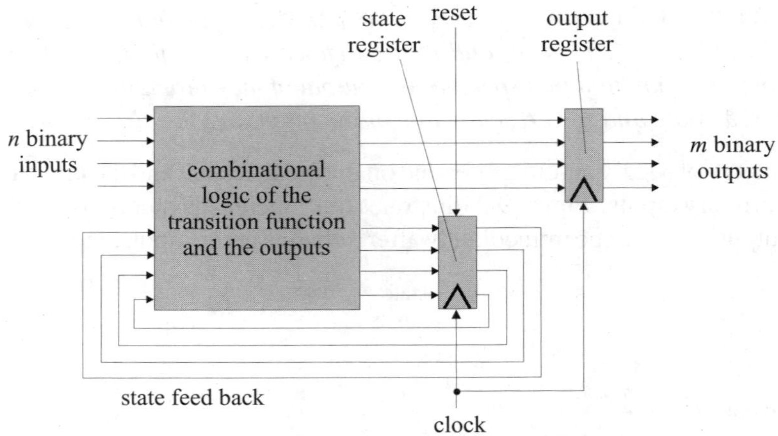

Figure 9.3. Synchronous Mealy machine [70]

and

$$\lambda^+(s,w) = \begin{cases} \lambda(s,w) & \text{, if } |w| = 1 \\ (\lambda^+(s, w_{1\ldots|w|-1}), \lambda(\delta^+(s, w_{1\ldots|w|-1}), w_{|w|})) & \text{, if } |w| \geq 2 \end{cases},$$

respectively.

For a subset $T \subseteq S$ of the states, the set of the states of S which can be reached from states of T is denoted by $REACH_M(T)$, or $REACH(T)$ in short, which is formally defined as

$$REACH(T) = T \cup \{s \in S;\ \exists t \in T\ \exists w \in I^+ :\ \delta^+(t,w) = s\}.$$

Furthermore, let $PRED(T)$ and $SUCC(T)$ denote the set of the immediate predecessors and successors, respectively, of states of T, i.e.,

$$PRED(T) = \{s \in S;\ \exists t \in T\ \exists w \in I :\ \delta(s,w) = t\}$$

and

$$SUCC(T) = \{s \in S;\ \exists t \in T\ \exists w \in I :\ \delta(t,w) = s\}.$$

1.2 Representations of Finite State Machines

Basically there are two different categories of FSM representations – explicit and implicit representations.

1.2.1 Explicit representations

The traditional approach to represent an FSM $M = (S, I, O, \delta, \lambda, r)$ is to list for each state s and each input symbol $w \in I$ the next state $\delta(s,w)$ and the output symbol $\lambda(s, w)$. Since the states, input symbols, and output symbols are encoded, the transition function δ and the output function λ are (incompletely specified) Boolean functions,

$$\delta : \{0,1\}^q \times \{0,1\}^n \rightarrow \{0,1\}^q$$

and

$$\lambda : \{0,1\}^q \times \{0,1\}^n \rightarrow \{0,1\}^m.$$

Thus, they can be represented as truth tables (see Chapter 3 Section 2.1).

A more pictorial representation of an FSM is the *state transition diagram* which is a directed graph $G = (S, E, label)$ whose nodes are given by the set S of the states of the FSM. There is an edge $e \in E$ from state s to state t labelled with $w|y$, i.e, $label(e) = w|y$, if and only if there is an input symbol $w \in I$ such that $\delta(s, w) = t$ and $\lambda(s, w) = y$ hold. For illustration, see the state transition diagram shown in Figure 9.4 (where we have not encoded the states, for the sake of simplicity). In this figure, the following finite state machine is

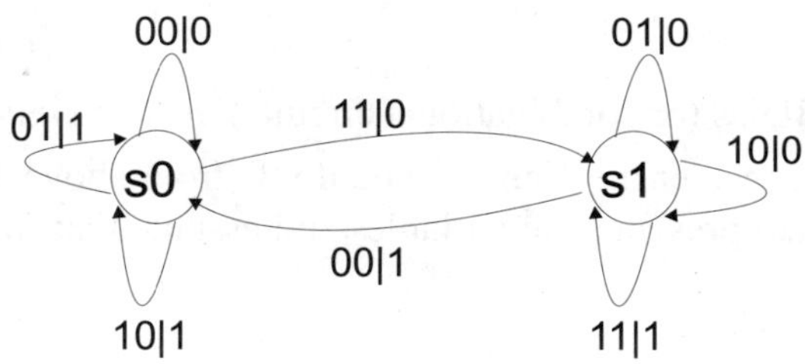

Figure 9.4. Serial Adder [17]

represented: $M = (\{s0, s1\}, \{0,1\}^2, \{0,1\}, \delta, \lambda, s_0)$ with

$$\delta(s, w) = \begin{cases} s0 & \text{, if } s = s0 \text{ and } w \in \{00, 01, 10\} \\ s1 & \text{, if } s = s0 \text{ and } w \in \{11\} \\ s1 & \text{, if } s = s1 \text{ and } w \in \{01, 10, 11\} \\ s0 & \text{, if } s = s1 \text{ and } w \in \{00\} \end{cases}$$

and

$$\lambda(s, w) = \begin{cases} 0 & \text{, if } s = s0 \text{ and } w \in \{00, 11\} \\ 1 & \text{, if } s = s0 \text{ and } w \in \{01, 10\} \\ 0 & \text{, if } s = s1 \text{ and } w \in \{01, 10\} \\ 1 & \text{, if } s = s1 \text{ and } w \in \{00, 11\}. \end{cases}$$

It describes a serial adder. In the ith clock cycle, the FSM is in state $s0$, if the carry-bit c_{i-1} of position $i-1$ was 0; otherwise it is in state $s1$. Note, that $s0$ is the start state as well. The ith bit a_i and b_i of the two operands a and b are read. The corresponding sum-bit which is $c_{i-1} \oplus a_i \oplus b_i$ determines the output. The carry-bit c_i which is $(a_i \cdot b_i) + ((a_i \oplus b_i) \cdot c_{i-1})$ determines the transition.

Unfortunately, it can be difficult to describe complex FSMs in this way. Just, consider a counter modulo 2^q. This sequential circuit consists of 2^q states. Of course, it cannot be represented by an explicit representation whenever q is large.

1.2.2 Implicit representations

In contrast to the tabular form, an implicit representation of an FSM does not explicitly enumerate the states. The basic idea is to represent the transition function

$$\delta : \{0,1\}^{q+n} \to \{0,1\}^q$$

and the output function

$$\lambda : \{0,1\}^{q+n} \to \{0,1\}^m$$

by q and m ROBDDs (or combinational circuits), respectively.

For illustration, consider the counter modulo 2^q from above, once again. Let $s_{q-1}, \ldots, s_0$ be the present state variables and assume that for all $c \in \mathbb{N}_0$, the equation

$$c \mod 2^q = \sum_{i=0}^{q-1} s_i 2^i$$

holds after c clock cycles. It is easy to see that the transition function can be represented by q ROBDDs $\Delta_{q-1}, \ldots, \Delta_0$ of linear size, i.e., $|\Delta_i| = \mathcal{O}(q)$, such that Δ_i determines the ith bit-position of the next state. Figure 9.5 shows ROBDD Δ_i.

This proves that large FSM can be compactly represented by ROBDDs. Furthermore, remember that we have shown in Chapter 3, that ROBDDs are particularly suitable for both, analysis and synthesis of Boolean functions.

2. Equivalence of Finite State Machines

Before going into details, we have to formally define the equivalence of two FSMs. Intuitively, two finite state machines $M_1 = (S_1, I_1, O_1, \delta_1, \lambda_1, r_1)$ and $M_2 = (S_2, I_2, O_2, \delta_2, \lambda_2, r_2)$ are equivalent if they have the same input/output behavior, i.e., $I_1 = I_2$, $O_1 = O_2$, and for an arbitrary input sequence $w \in I_1^+$

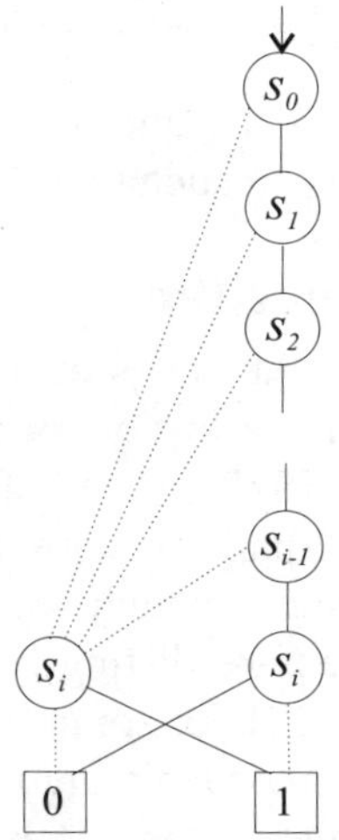

Figure 9.5. ROBDD Δ_i of the counter modulo 2^q

applied to both start states r_1 and r_2 the same output sequence results. To formalize this notion, we have to introduce equivalent states.

DEFINITION 9.3 (FUNCTIONALLY EQUIVALENT STATES)
Two states $s_1 \in S_1$ and $s_2 \in S_2$ are functionally equivalent, written $s_1 \sim s_2$ if and only if

$$\forall w \in I_1^+ : \lambda_1^+(s_1, w) = \lambda_2^+(s_2, w).$$

Otherwise, we say that s_1 and s_2 are distinguishable and write $s_1 \not\sim s_2$.

DEFINITION 9.4 (FUNCTIONALLY EQUIVALENT MACHINES)
The finite state machines M_1 and M_2 are called functionally equivalent, written $M_1 \sim M_2$, if the start states r_1 and r_2 are functionally equivalent, i.e., $r_1 \sim r_2$.

Although Definition 9.3 is given for two different automata, it can be used for a single FSM to find equivalent states. To ease the presentation of basic algorithms for deciding whether two states are equivalent, we introduce a further notion.

DEFINITION 9.5 (C-EQUIVALENT STATES)
Two states $s_1 \in S_1$ and $s_2 \in S_2$ are c-equivalent for some constant $c \in \mathbb{N}$, written $s_1 \sim_c s_2$, if

$$\forall w \in I_1^+ \text{ with } |w| \leq c : \lambda_1^+(s_1, w) = \lambda_2^+(s_2, w).$$

Otherwise, we say that s_1 *and* s_2 *are c-distinguishable and we write* $s_1 \not\sim_c s_2$.

It is obvious, that two states $s_1 \in S_1$ and $s_2 \in S_2$ are distinguishable if and only if there exists a constant $c \in \mathbb{N}$ such that s_1 and s_2 are c-distinguishable.

2.1 The table-filling algorithm

The basic algorithm for finding the states of an FSM $M = (S, I, O, \delta, \lambda, r)$ which are equivalent is a recursive discovery of distinguishable pairs. It is known as *table filling algorithm* [62]. Instead of computing pairs of equivalent states, it concentrates on finding pairs of states that are distinguishable. In an initialization step, all pairs of states which are 1-distinguishable are marked. As noticed, these pairs of states are distinguishable. Then, we are looking for pairs of states p and q such that for some input value $w \in I$, $\delta(p, w)$ and $\delta(q, w)$ are a pair of states known to be distinguishable. It is easy to see, that in this case p and q are distinguishable, too. They are marked by the algorithm. The algorithm stops if no such pair can be found, any more. A sketch of the algorithm is shown in Figure 9.6.

The algorithm computes the set E of pairs of states which are equivalent since any pair of states which is not marked by the algorithm is functionally equivalent.

THEOREM 9.1 ([62])
If two states are not distinguishable by the table-filling algorithm, then the states are equivalent.

PROOF: Assume that there is a pair of states which is distinguishable although the table-filling algorithm does not find it. We call this pair a bad pair.

Because of the initialization step of the table-filling algorithm, the algorithm marks all pairs of states which are 1-distinguishable.

Now, let $w \in I^+$ be the shortest sequence which distinguishes a bad pair. Let $\{p, q\}$ be a bad pair with

$$\lambda^+(p, w) \neq \lambda^+(q, w).$$

Because of the initialization step of the table-filling algorithm, the states of bad pairs are 1-equivalent. Thus the length $|w|$ of sequence w has to be larger than 1 and it follows for $s = \delta(p, w_1)$ and $t = \delta(q, w_1)$

$$\begin{aligned}
\lambda^+(s, w_{2\ldots|w|}) &= \lambda^+(\delta(p, w_1), w_{2\ldots|w|}) \\
&= \lambda^+(p, w)_{2\ldots|w|} \\
&\neq \lambda^+(q, w)_{2\ldots|w|} \quad \text{since } \lambda^+(p, w) \neq \lambda^+(q, w) \text{ but } p \sim_1 q \\
&= \lambda^+(\delta(q, w_1), w_{2\ldots|w|}) \\
&= \lambda^+(t, w_{2\ldots|w|}).
\end{aligned}$$

```
TABLE_FILLING(M)
begin

  E = {{p, q}; p, q ∈ S and p ~1 q};
  // E contains the pairs of states that are 1-equivalent
  D = {{p, q}; p, q ∈ S} \ E;
  // D contains the pairs of states found to be distinguishable

  repeat
    found1 = 0;
    foreach {p, q} ∈ E do
      found2 = 0;
      foreach w ∈ I do
        if ({δ(p, w), δ(q, w)} ∈ D)
        then E = E \ {{p, q}};
             D = D ∪ {{p, q}};
             found1 = found2 = 1;
        fi;
      od until found2==1;
    od;
  until found1==0;

  return E;

end;
```

Figure 9.6. Table-filling algorithm

That is, s and t are distinguishable by a sequence which is smaller than w. Because of the choice of w, the pair $\{s, t\}$ is found to be distinguishable by the table-filling algorithm and the table-filling algorithm does not stop until the pair $\{p, q\}$ is marked as well. This is a contraction to our assumption that $\{p, q\}$ is a bad pair. ∎

Let us now check the equivalence of two FSMs $M_1 = (S_1, I, O, \delta_1, \lambda_1, r_1)$ and $M_2 = (S_2, I, O, \delta_2, \lambda_2, r_2)$. Therefore, let us construct the FSM $M_1 \cup M_2$ which is the "disjoint union of M_1 and M_2". Let us assume without loss of generality that $S_1 \subseteq \{0,1\}^q$, $S_2 \subseteq \{0,1\}^q$, and $O \subseteq \{0,1\}^m$ for some $q, m \in \mathbb{N}$. Then, formally the FSM $M_1 \cup M_2$ can be defined as

$$M_1 \cup M_2 = (\overline{S_1} \cup \overline{S_2} \cup \{\rho\}, I, \overline{O} \cup \{\tau\}, \delta, \lambda, \rho)$$

with

- $$\begin{aligned} \overline{S_1} &= \{0\} \times \{1\} \times S_1 \\ \overline{S_2} &= \{1\} \times \{0\} \times S_2 \\ \rho &= (0, \ldots, 0) \in \{0,1\}^{q+2} \end{aligned}$$

- $$\begin{aligned} \overline{O} &= \{1\} \times O \\ \tau &= (0, \ldots, 0) \in \{0,1\}^{m+1} \end{aligned}$$

- $$\delta(s, w) = \begin{cases} (0, 1, \delta_1(s_{3 \ldots q+2}, w)) & \text{, if } (s_1, s_2) = (0, 1) \\ (1, 0, \delta_2(s_{3 \ldots q+2}, w)) & \text{, if } (s_1, s_2) = (1, 0) \\ \rho & \text{, if } s = \rho \end{cases}$$

- $$\lambda(s, w) = \begin{cases} (1, \lambda_1(s_{3 \ldots q+2}, w)) & \text{, if } (s_1, s_2) = (0, 1) \\ (1, \lambda_2(s_{3 \ldots q+2}, w)) & \text{, if } (s_1, s_2) = (1, 0) \\ \tau & \text{, if } s = \rho. \end{cases}$$

That is, the encodings of the states of the FSM M_1 have sequence $(0, 1)$ as prefix and the encodings of the states of FSM M_2 have sequence $(1, 0)$ as prefix. In some sense, $M_1 \cup M_2$ is absurd since the FSM remains in the start state ρ independently of the input sequence applied. However, applying the table filling algorithm to $M_1 \cup M_2$ results in the computation of all pairs of states which are equivalent. Thus, M_1 and M_2 are functionally equivalent FSMs if and only if the pair $((0, 1, r_1), (1, 0, r_2))$ is found equivalent by the table-filling algorithm.

2.2 Product machine

Another approach which is similar to the table-filling algorithm is based on the so-called product machine, or product automaton, of two FSMs. In fact, it is the algorithm used by today's verification engineers.

DEFINITION 9.6 (PRODUCT MACHINE) *Given two finite state machines M_1 and M_2 with $M_i = (S_i, I, O, \delta_i, \lambda_i, r_i)$ for $i = 1, 2$. The product machine*

$$M_1 \times M_2 = (S_{M_1 \times M_2}, I, O_{M_1 \times M_2}, \delta_{M_1 \times M_2}, \lambda_{M_1 \times M_2}, r_{M_1 \times M_2})$$

of M_1 and M_2 is an FSM with

- $S_{M_1 \times M_2} = S_1 \times S_2$,
- $O_{M_1 \times M_2} = \{0, 1\}$,
- $\delta_{M_1 \times M_2}((s_1, s_2), w) = (\delta_1(s_1, w), \delta_2(s_2, w))$,
- $\lambda_{M_1 \times M_2}((s_1, s_2), w) = 1 \iff \lambda_1(s_1, w) = \lambda_2(s, w)$, *and*
- $r_{M_1 \times M_2} = (r_1, r_2)$.

A state s of $S_{M_1 \times M_2}$ is called differentiating state if there exists an input symbol $w \in I$ such that $\lambda(s, w) = 0$. Furthermore, a sequence $w \in I^+$ of input symbols is called distinguishing sequence for s if $\lambda^+(s, w) = 0$.

Obviously, FSMs M_1 and M_2 are functionally equivalent if and only if no differentiating state $s \in S_{M_1 \times M_2}$ can be reached from start state $r_{M_1 \times M_2}$ in the product machine $M_1 \times M_2$, i.e.,

THEOREM 9.2 *Two FSMs M_1 and M_2 are equivalent if and only if*

$$\forall s \in REACH_{M_1 \times M_2}(\{(r_1, r_2)\})\ \forall w \in I : \lambda_{M_1 \times M_2}(s, w) = 1.$$

In order to decide equivalence, a traversal needs to be performed on the product state space. We are going to focus upon this subject in the next section.

3. State space traversal

Equivalence checking of FSMs, typically proceeds by traversing the state space. In this section, we will concentrate on this problem and propose the basic algorithms for efficient state space traversal.

3.1 The basic approaches

There are two different approaches for state space traversal, forward and backward traversal. Both algorithms are sketched in Figure 9.7 and Figure 9.8. They refer to the approach proposed by Theorem 9.2.

3.1.1 Forward state space traversal

Mathematically, *reached state computation* which is the process of taking the start state r and finding all states reachable from this state, i.e., all states s such that there is a sequence of input symbols which leads the FSM into s from the start state, is performed by a fixed point calculation [32, 122]:

$$REACH_0 = \{r\}$$

$$REACH_{k+1} = REACH_k \cup SUCC(REACH_k).$$

The algorithm of Figure 9.7 is based on this principle. It realizes a *forward traversal* of the product automaton.

FORWARD_TRAVERSAL(M)
// M is the product automaton of two FSMs M_1 and M_2
begin

$REACH = \{r\}$;
// $REACH$ contains the state visited by the algorithm

$FRONT = REACH$;
// $FRONT$ contains those states of $REACH$ which
// have not yet been tested whether new states are reachable

repeat
if ($\exists s \in FRONT\, \exists w \in I:\ \lambda(s, w) == 0$)
then return ´´The FSMs are not equivalent´´;
fi;

$FRONT = SUCC(FRONT) \setminus REACH$;
$REACH = REACH \cup FRONT$;
until $FRONT$==$\emptyset$;

return ´´Both FSMs are equivalent´´;

end;

Figure 9.7. Sketch of the equivalence checking of FSMs by forward state space traversal.

During the initialization phase, variable $REACH$ is set to $\{r\}$. Variable $FRONT$ contains those states of variable $REACH$ whose successors have not yet been visited.

In each iteration, the algorithm determines $SUCC(FRONT)$. The new front is given by these successors but the states already contained in REACH. If the algorithm encounters a differentiating state s, i.e., $\lambda(s, w) = 0$ for some input value $w \in I$, it reports that the FSMs under verification are not equivalent and stops (see Theorem 9.2). If the assignment of $REACH$ is stable, i.e., $REACH = SUCC(REACH)$ which is equivalent to $FRONT = \emptyset$, then all the reachable states have been visited and both FSMs are equivalent since no differentiating state could be reached from start state r.

3.1.2 Backward state space traversal

Backward traversal moves through the state space in the opposite direction. It

```
BACKWARD_TRAVERSAL(M)
// M is the product automaton of two FSMs M1 and M2
begin

  FRONT = {s ∈ S; ∃w ∈ I : λ(s, w) == 0};
  EQUIV = S \ FRONT;

  repeat
    if (r ∈ FRONT)
    then return ´´The FSMs are not equivalent´´;
    fi;

    FRONT = PRED(FRONT) ∩ EQUIV;
    EQUIV = EQUIV \ FRONT;
  until FRONT==∅;

  return ´´Both FSMs are equivalent´´;

end;
```

Figure 9.8. Sketch of the equivalence checking of FSMs by backward state space traversal.

starts with set $FRONT$ which now contains the differentiating states of the product machine. Then, it iteratively computes the predecessors of these differentiating states. If during backward traversal the start state r is encountered, a differentiating state can be reached from r and both FSMs are not equivalent. Note that in each iteration, the new front is obtained by intersecting the set of the predecessors of FRONT and the set EQUIV which contains the states which have not been in the front, yet. This is motivated by the fact that the FSMs under verification are not equivalent if and only if there is a simple path $(s_0, \ldots, s_j)$ with

- $s_0 = r$,
- $\forall i \in \{0, \ldots, j-1\} : s_{i+1} \in SUCC(\{s_i\})$,
- $\forall i \in \{0, \ldots, j-1\} : s_i$ is a non-differentiating state, and
- s_j is a differentiating state.

3.2 ROBDD based implementation

The crucial operation of forward and backward state space traversal is the computation of the set of successor states and predecessor states, respectively. Both operations can be efficiently executed by using implicit representations of sets of states.

DEFINITION 9.7 (IMPLICIT REPRESENTATION OF SET OF STATES) *Let $T \subseteq \{0,1\}^q$ be a subset of states. Then subset T is uniquely represented by the characteristic function $\chi_T : \{0,1\}^q \to \{0,1\}$ defined by*

$$\forall s \in \{0,1\}^q : \chi_T(s) = 1 \iff s \in T.$$

Analogously, the transition function δ of an FSM can be represented as a Boolean function as well. We just have to take the characteristic function of the state transition relation.

DEFINITION 9.8 (STATE TRANSITION RELATION) *Given a finite state machine $M = (S, I, O, \delta, \lambda, r)$, the state transition relation $\Delta \subseteq S \times I \times S$ of M is defined as*

$$\forall s,t \in \{0,1\}^q \; \forall w \in \{0,1\}^n : (s,w,t) \in \Delta \iff \delta(s,w) = t.$$

The characteristic function χ_Δ of Δ is a Boolean function of $\mathcal{B}_{2q+n}$ defined on $2q+n$ variables, $\sigma_1, \ldots, \sigma_q$, $\omega_1, \ldots, \omega_n$, and $\tau_1, \ldots, \tau_q$, which we call *present state variables*, *input variables*, and *next state variables*, respectively. For illustration, we can consider the present state variables as the outputs and the next state variables as the inputs of the state register (see Figure 9.3).

3.2.1 Implementation of forward traversal

With the new definitions let us consider the algorithm for equivalence checking of FSMs by forward state space traversal again.

Let $s, t \in \{0,1\}^q$ be two states, T be a subset of states, and $w \in \{0,1\}^n$ be an assignment to the input variables $\omega_1, \ldots, \omega_n$. Now, state s is element of T and the FSM moves from s to t under the assignment w if and only if the expression

$$\chi_T(s) \cdot \chi_\Delta(s,w,t)$$

evaluates to 1. Thus the characteristic function of

$$\{(s,w,t) \in \{0,1\}^{2q+n};\; s \in T \text{ and } \delta(s,w) = t\}$$

is given by

$$\chi_T(\sigma_1, \ldots, \sigma_q) \cdot \chi_\Delta(\sigma_1, \ldots, \sigma_q, \omega_1, \ldots, \omega_n, \tau_1, \ldots, \tau_q).$$

```
FORWARD_TRAVERSAL(M)
begin

  χ_REACH = χ_{r};
  χ_FRONT = χ_REACH;

  repeat
    if ((χ_FRONT(σ) · NOT(λ(σ,ω))) ≠ 0)
    then report ´´The FSMs are not equivalent´´;
    fi;

    χ_SUCC = (∃ω, σ : χ_FRONT(σ) · χ_Δ(σ, ω, τ));
    χ_FRONT = χ_SUCC · NOT(χ_REACH);
    χ_REACH = χ_REACH + χ_FRONT;

  until χ_FRONT==0;

  report ´´Both FSMs are equivalent´´;

end;
```

Figure 9.9. Equivalence Checking of FSMs by forward state space traversal.

We obtain the set $\{(s,t);\ s \in T \text{ and } t \in SUCC(s)\}$ by performing an existential quantification over the input variables, that is

$$(\exists \omega_1, \ldots, \omega_n : \chi_T(\sigma_1, \ldots, \sigma_q) \cdot \chi_\Delta(\sigma_1, \ldots, \sigma_q, \omega_1, \ldots, \omega_n, \tau_1, \ldots, \tau_q))$$

or

$$(\exists \omega : \chi_T(\sigma) \cdot \chi_\Delta(\sigma, \omega, \tau)),$$

in short. By further performing an existential quantification over the present state variables, we obtain the characteristic function of $SUCC(T)$, that is

$$\chi_{SUCC(T)}(\tau) = (\exists \sigma : (\exists \omega : \chi_T(\sigma) \cdot \chi_\Delta(\sigma, \omega, \tau))).$$

Thus if both, the Boolean function χ_T and the Boolean function χ_Δ are given by ROBDDs, then the ROBDD of the characteristic function $\chi_{SUCC(T)}$ of $SUCC(T)$ can be computed by performing the AND of the given ROBDDs and an existential quantification over the input variables and the present state variables. Both operations, the logical-and operator and the existential quantification which is performed $q + n$ times can be executed in polynomial time as shown in Chapter 3.

Finally, the branch condition $(\exists s \in FRONT\, \exists w \in I : \lambda(s,w) == 0)$ can be realized as

$$(\chi_{FRONT}(\sigma) \cdot NOT(\lambda(\sigma,\omega))) \neq \mathbf{0}.$$

All these considerations result in the algorithm shown in Figure 9.9. Note that, in order to be accurate, the statement

$$\chi_{FRONT} = (\exists \omega, \sigma : \chi_{FRONT}(\sigma) \cdot \chi_{\Delta}(\sigma,\omega,\tau));$$

has to be replaced by

$$\chi_{FRONT} = (\exists \omega, \sigma : \chi_{FRONT}(\sigma) \cdot \chi_{\Delta}(\sigma,\omega,\tau))_{\tau=\sigma};$$

in each iteration, i.e., the variables $\tau_1, \ldots, \tau_q$ have to be substituted by the variables $\sigma_1, \ldots, \sigma_q$.

3.2.2 Implementation of backward traversal

Analogously, the backward state space traversal can be implemented. The implementation is shown in Figure 9.10. According to the considerations from

---

BACKWARD_TRAVERSAL(M)
begin

$\chi_{FRONT} = (\exists \omega : NOT(\lambda(\sigma,\omega)))$;
$\chi_{EQUIV} = \chi_S \cdot NOT(\chi_{FRONT})$;

repeat
if $((\chi_{\{r\}} \cdot \chi_{FRONT}) \neq \mathbf{0})$
then report ´´The FSMs are not equivalent´´;
fi;

$\chi_{PRED} = (\exists \omega, \tau : \chi_{FRONT}(\tau) \cdot \chi_{\Delta}(\sigma,\omega,\tau))$;
$\chi_{FRONT} = \chi_{PRED} \cdot \chi_{EQUIV}$;
$\chi_{EQUIV} = \chi_{EQUIV} \cdot NOT(\chi_{FRONT})$;
until χ_{FRONT}==**0**;

report ´´Both FSMs are equivalent´´;

end;

Figure 9.10. Equivalence checking of FSMs by backward state space traversal.

above, it is easy to prove that

$$\chi_{PRED(T)}(\sigma) = (\exists \omega, \tau : \chi_T(\tau) \cdot \chi_{\Delta}(\sigma,\omega,\tau))$$

holds. The implementation of the remaining operations of the algorithm shown in Figure 9.8 is straightforward as well.

3.3 Error trace generation

When the FSMs under verification are not equivalent, a simple negative answer of the verification algorithm is unsatisfactory. It is usually desirable to derive an input sequence that distinguishes the two FSMs. This process is called *error trace generation.*

Without loss of generality, let us concentrate on the forward state space traversal algorithm. We have to extend the algorithm such that distinguishing input sequences can be generated. For this purpose we have to store the front of each iteration cycle separately. Figure 9.11 shows the extended algorithm.

Now, error trace generation is rather easy. Assume that the procedure ERROR_TRACE_GENERATION is called with the parameters $i \in \mathbb{N}$ and $error_i \in S$, i.e., $error_i \in FRONT[i]$. First, we have to look for an input value w_i such that the distinguishing state $error_i$ actually outputs 0, i.e., $\lambda(error_i, w_i) = 0$. Input value w_i is the last element of the distinguishing input sequence which is generated. The next step is to find a predecessor state of $error_i$ which is closer to start state r. For this purpose, just compute

$$PRED(\{error_i\}) \cap FRONT[i-1]$$

and pick a state $error_{i-1}$ from this intersection. The $(i-1)th$ element of the distinguishing input sequence is set to some $w_{i-1} \in I$ such that

$$\delta(error_{i-1}, w_{i-1}) = error_i.$$

In the same manner, the procedure computes the remaining (inverse) path $error_{i-2}, \ldots, error_1$ and the corresponding input values $w_{i-2}, \ldots, w_1 \in I$. Since $FRONT[1]$ only consists of start state r, state $error_1$ is the start state and $w_1, \ldots, w_i$ is a distinguishing input sequence.

```
FORWARD_TRAVERSAL_WITH_ERROR_TRACE_GENERATION(M)
begin

  i = 0;
  REACH = {r};
  FRONT[1] = REACH;
  // FRONT[i] is the front of the ith iteration

   repeat
    i++;
    if (∃ s ∈ FRONT[i] ∃w ∈ I : λ(s, w) == 0)
    then ERROR_TRACE_GENERATION(M,i,s);
    fi;

    FRONT[i+1] = SUCC(FRONT[i]) \ REACH;
    REACH = REACH ∪ FRONT[i+1];
   until FRONT[i+1]==∅;

  report ´´Both FSMs are equivalent´´;

end;


ERROR_TRACE_GENERATION(M,i,error)
begin

  w = [ PICK_ONE_ELEMENT_OF({v ∈ I; λ(error, v) = 0}) ];

  for j = i − 1 downto 1
  do
   pred = PICK_ONE_ELEMENT_OF(PRED({error}) ∩ FRONT[j]);
   w = [ PICK_ONE_ELEMENT_OF({v ∈ I; δ(pred, v) = error}) ] ∘w;
   // ´´∘´´ denotes the concatenation of finite sequences

   error = pred;
  od;

  return w;

end;
```

Figure 9.11. Forward state space traversal with error trace generation

Chapter 10

THE LATCH CORRESPONDENCE PROBLEM

This chapter addresses the problem of establishing the unknown correspondence of the latches (memory elements) of two sequential circuits which have the same encoding of their states. When circuit modifications do not alter the state encoding, this *Latch Correspondence Problem* has to be solved in order to reduce the sequential equivalence checking problem to an equivalence check of combinational circuits. If a correspondence of the latches in the two sequential circuits can be established, then it is possible to break the circuits into corresponding combinational blocks (cf. Figure 9.1 and Figure 9.3) and the verification problem reduces to an equivalence check of the combinational circuits defined by the latch boundaries [77, 117]. In this case, the sequential circuits to be considered are functionally equivalent if the combinational blocks are functionally equivalent.

There are several cases in which the correspondence may exist but is unknown. For example, existing commercial tools doing design transformations, such as synthesis or clock tree insertion, often do not preserve signal names. So it becomes impossible to establish a latch correspondence without further help. This especially happens when a hierarchically specified design is being flattened as part of the transformation.

Several approaches to compute latch correspondences have been proposed in literature [24, 141, 107]. In this chapter we concentrate on the approach presented in [107] which can be easily included whenever ROBDDs (see Definition 3.6 on page 31) are used as the representation form in the verification tool. Experimental runs applied to a large set of benchmarks have proven the efficiency of the approach [107].

The combinational equivalence check on the primary output and the transition functions can be done using ROBDDs as shown in Chapter 4. Since ROBDDs are a canonical form for representing Boolean functions for fixed variable order, equivalence check reduces to checking if the two canonical forms are the same [15, 20]. However, without the correspondence of the latches the ROBDDs of the combinational blocks of the two finite state machines under consideration could be ordered with respect to different variable orders and we cannot simply check whether both ROBDDs are the same. Thus, we need to establish this correspondence first. Given k latches, there are $k!$ possible correspondences. This rules out any exhaustive algorithms for establishing the correspondence for circuits encountered in practice.

There is a similar problem in combinational logic verification which we have discussed in Chapter 8, the permutation independent Boolean comparison. The new problem does not only resemble the permutation independent Boolean comparison but there is a similar approach to solve it as well. Again, signature-based methods can be successfully applied. These methods can be especially used as an enhancement to commercial tools solving the problem, i.e., it can be reserved for tough cases where those tools fail. It can be applied to those pairs of sequential circuits where a direct correspondence between the latches is possible, i.e., there is the same state encoding.

The application is straightforward, namely derive signatures for each latch in order to uniquely identify this latch. In [121], the authors have used this method to identify corresponding latches in order to be able to simplify the product machine traversal for sequential verification. There, they have used input signatures and have made the observation that sometimes the task still remains too complex. The ideas presented in [107] are able to improve these results. They do not just combine the input signatures known from literature, but also develop novel signatures which are especially suited for the latch equivalence problem. These signatures give better results, i.e., the set of pairs of latches which are candidates for correspondence is determined more precisely.

The chapter starts with a problem description in Section 1. Section 2 describes the differences between establishing unknown input variable correspondence and establishing unknown latch correspondence, introduces special signature functions for latches which can be easily computed on ROBDDs, and explains the solution paradigm which uses these signature functions on an example.

1. Problem description

For illustration, we refer to Figure 10.1a and Figure 10.1b.

Let $\mathcal{F}_{n,m,k}$ be the set of the finite state machines (FSM) with n primary inputs

$$X = [x_1, x_2, \ldots, x_n],$$

m primary outputs

$$Y = [y_1, y_2, \ldots, y_m],$$

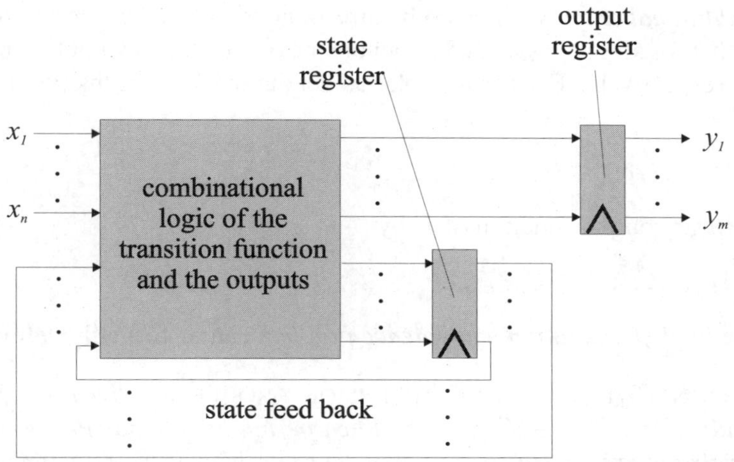

Figure 10.1a. A synchronous Mealy machine F

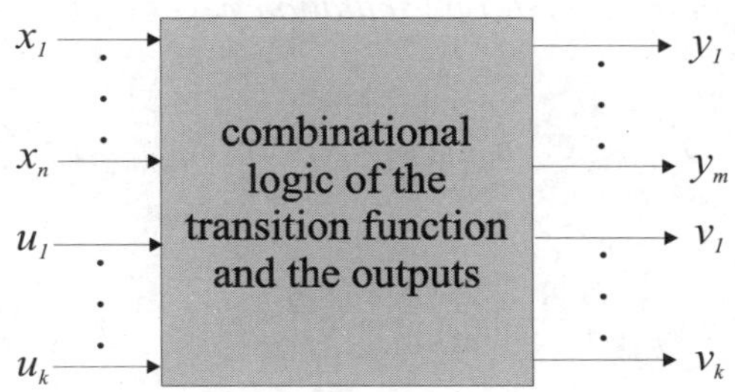

Figure 10.1b. The combinational block $Com(F)$.

and k latches

$$L = [l_1, l_2, \dots, l_k].$$

For our purpose, we can concentrate on the combinational blocks of the sequential circuits in $\mathcal{F}_{n,m,k}$. The combinational part $Com(F)$ of a sequential circuit $F \in \mathcal{F}_{n,m,k}$ can be regarded as a Boolean function with $n + k$ primary inputs and $m + k$ primary outputs. Let

$$U = [u_1, u_2, \dots, u_k]$$

be the *additional inputs* and let

$$V = [v_1, v_2, \dots, v_k]$$

be the *additional outputs* which both correspond to the latches "removed". We assume that for all $j \in \mathbb{N}_k$ u_j and v_j denote the output and the input of the same latch l_j, respectively. Furthermore, let us denote the k-ary transition function of F by

$$\delta = [\delta_1, \ldots, \delta_k] \in \mathcal{B}_{n+k,k}$$

and the m-ary output function of F by

$$\lambda = [\lambda_1, \ldots, \lambda_m] \in \mathcal{B}_{n+k,m}.$$

Now, the *latch permutation equivalence problem* can be formally defined.

DEFINITION 10.1 (LATCH PERMUTATION PROBLEM) *Given two sequential circuits $F^{(1)}, F^{(2)} \in \mathcal{F}_{n,m,k}$, the latch permutation equivalence problem which is also referred as latch correspondence problem is the decision problem as to whether a correspondence π between the latches of $F^{(1)}$ and $F^{(2)}$ exists, such that the two synchronous sequential circuits $F^{(1)}$ and $F^{(2)}$ have their combinational parts functionally equivalent using this correspondence. More formally, the problem is to find a permutation $\pi \in \mathcal{P}er(\mathbb{N}_k)$ such that for all $j \in \mathbb{N}_m$*

$$\lambda_j^{(1)}(x_1, \ldots, x_n, u_{\pi(1)}^{(1)}, \ldots, u_{\pi(k)}^{(1)}) = \lambda_j^{(2)}(x_1, \ldots, x_n, u_1^{(2)}, \ldots, u_k^{(2)})$$

and for all $j \in \mathbb{N}_k$

$$\delta_{\pi(j)}^{(1)}(x_1, \ldots, x_n, u_{\pi(1)}^{(1)}, \ldots, u_{\pi(k)}^{(1)}) = \delta_j^{(2)}(x_1, \ldots, x_n, u_1^{(2)}, \ldots, u_k^{(2)})$$

hold. (For the notations, we refer to Chapter 8 Section 1.)

Note that the additional inputs $u_1^{(1)}, \ldots, u_k^{(1)}$ of the combinational part of $F^{(1)}$ have to be permutated in the exact same manner as the additional outputs $v_1^{(1)}, \ldots, v_k^{(1)}$ (the assignment of $v_j^{(1)}$ is determined by $\delta_j^{(1)}$) since latches are permutated and $u_j^{(1)}$ and $v_j^{(1)}$ are the output and the input of the same latch. Furthermore, note that we assume without loss of generality that the correspondences between the primary inputs and primary outputs of $F^{(1)}$ and $F^{(2)}$ are known. If these correspondences are unknown as well, then just apply the methods discussed in Chapter 8.

The similarity of the latch correspondence problem to the combinational permutation equivalence problem is obvious. We need to find a unique and permutation independent description for each latch. The difference is that each latch has to be considered as input *and* as output of the combinational part of the sequential circuit and that these inputs are permutated in exactly the same manner as the outputs.

2. Signatures

Let us remember the problem of finding a permutation independent correspondence between the input variables of Boolean functions of $\mathcal{B}_n$ first. For this, let $f \in \mathcal{B}_n$. In Chapter 8, we have defined a *signature* as a description of an input variable of f which is independent of any permutation of the (remaining) inputs of a Boolean function f. Applying signatures helps solving the combinational permutation equivalence problem by identifying each input variable independently of permutation. Any possible correspondence between the input variables of two Boolean functions is restricted to a correspondence between variables with the same signature. So, if each variable of a Boolean function f has a unique signature, then there is at most one possible correspondence to the variables of any other Boolean function.

This concept can be used to attack the *latch correspondence problem* as well. Here we need to test whether there exists a correspondence between the *latches* of two synchronous sequential circuits such that their combinational logic is functionally equivalent under this correspondence.

Let us characterize this new problem and underline the differences from the input correspondence problem. As explained above, we assume that the correspondence between the primary inputs of the two sequential circuits is known. So we have the first piece of additional information. Furthermore, we assume that we know the correspondence between the primary outputs $y_1, \ldots, y_m$ of the sequential circuits under consideration. These primary outputs represent the output functions $\lambda_1, \ldots, \lambda_m$ of the sequential circuit that depend on both, the primary inputs $x_1, \ldots, x_n$ of the sequential circuits and the additional inputs $u_1, \ldots, u_k$ which are due to the removal of the latches. So, we can use the primary outputs of the sequential circuits to compute input signatures for the latches. This is done in the exact same manner as for the combinational problem. We make some remarks regarding to these input signatures in Section 2.2. However, latches appear not only as inputs of the combinational part of the sequential circuit but as outputs as well. So, the most important difference between the combinational problem and the sequential problem is that we can use output signatures in addition to identify the latches independently of permutation. For instance, the satisfy count of the Boolean function δ_j realized by output v_j of the combinational part of the sequential circuit is such information on latch l_j. We will go in more details in Section 2.3.

In Section 2.4, we demonstrate how a unique possible correspondence for the latches can be established in practice. For illustration, we use the benchmark circuit *ex4.slif* from the benchmark set LGSynth91 [89].

2.1 Solution paradigm

Let us start with describing the solution paradigm on an example of two sequential circuits, $F^{(1)}$ and $F^{(2)}$, with four latch variables $[l_1^{(1)}, l_2^{(1)}, l_3^{(1)}, l_4^{(1)}]$ and $[l_1^{(2)}, l_2^{(2)}, l_3^{(2)}, l_4^{(2)}]$, respectively. Let the latches considered as inputs of the combinational part of the sequential circuits be denoted by $[u_1^{(1)}, u_2^{(1)}, u_3^{(1)}, u_4^{(1)}]$ and $[u_1^{(2)}, u_2^{(2)}, u_3^{(2)}, u_4^{(2)}]$, respectively. The corresponding outputs of the combinational blocks are $[v_1^{(1)}, v_2^{(1)}, v_3^{(1)}, v_4^{(1)}]$ and $[v_1^{(2)}, v_2^{(2)}, v_3^{(2)}, v_4^{(2)}]$, respectively.

At first, we look at the latches as inputs of the combinational block and compute an input signature $s_I(\lambda^{(1)}, u_i^{(1)})$ and $s_I(\lambda^{(2)}, u_i^{(2)})$ for $i = 1, 2, 3, 4$ with respect to the primary output functions $\lambda^{(1)}$ and $\lambda^{(2)}$ of $F^{(1)}$ and $F^{(2)}$, respectively. This is done in the exact same manner as described for the combinational permutation problem in Chapter 8. So let us suppose that we have two lists of input signatures that contain the same elements, namely

$$\mathcal{L}_I(F^{(1)}, [u_1^{(1)}, u_2^{(1)}, u_3^{(1)}, u_4^{(1)}]) = [2, 1, 3, 2]$$

and

$$\mathcal{L}_I(F^{(2)}, [u_1^{(2)}, u_2^{(2)}, u_3^{(2)}, u_4^{(2)}]) = [1, 2, 2, 3].$$

Now, based on these input signatures we can directly establish that any correspondence between the latches of $F^{(1)}$ and $F^{(2)}$ has to identify latch $l_2^{(1)}$ of $F^{(1)}$ with latch $l_1^{(2)}$ of $F^{(2)}$, and latch $l_3^{(1)}$ with latch $l_4^{(2)}$. Thus, a correspondence between the latches of $F^{(1)}$ and $F^{(2)}$ has been partially established. There is a possible correspondence between the latches of $F^{(1)}$ and $F^{(2)}$ for latch equivalence. However, both latches, $l_1^{(1)}$ and $l_4^{(1)}$ of $F^{(1)}$ could correspond to $l_2^{(2)}$ as well as to $l_3^{(2)}$ of $F^{(2)}$. In other words, aliasing occurs between the latches $l_1^{(1)}$ and $l_4^{(1)}$ of $F^{(1)}$ and between the latches $l_2^{(2)}$ and $l_3^{(2)}$ of $F^{(2)}$.

Since latches do not just correspond to inputs but to outputs of the combinational parts of the sequential circuits as well, we have a second possibility to distinguish between the latches with aliasing. We can use output signatures, for example, the satisfy counts of the corresponding transition functions, in order to try to uniquely identify them. In our example, we have to do this for the outputs $v_1^{(1)}$ and $v_4^{(1)}$ of $F^{(1)}$, and for $v_2^{(2)}$ and $v_3^{(2)}$ of $F^{(2)}$. So let us assume that

$$\mathcal{L}_O(F^{(1)}, [v_1^{(1)}, v_4^{(1)}]) = [4, 5]$$

and

$$\mathcal{L}_O(F^{(2)}, [v_2^{(2)}, v_3^{(2)}]) = [5, 4]$$

hold.

Now we are done, since we are able to establish a unique correspondence between the latches $l_1^{(1)}$ and $l_4^{(1)}$ of $F^{(1)}$ and the latches $l_2^{(2)}$ and $l_3^{(2)}$ of $F^{(2)}$ as well. Latch $l_1^{(1)}$ of $F^{(1)}$ has to correspond to latch $l_3^{(2)}$ of $F^{(2)}$, and latch $l_4^{(1)}$ of $F^{(1)}$ to latch $l_2^{(2)}$ of $F^{(2)}$. This gives us the only possible unique correspondence between the latches.

Furthermore, if we require that signatures have to be elements of an ordered set, we can establish a unique permutation independent order of the latches. In the case of our example, that would be the order $[l_2^{(1)}, l_1^{(1)}, l_4^{(1)}, l_3^{(1)}]$ for the latches of $F^{(1)}$ and the order $[l_1^{(2)}, l_3^{(2)}, l_2^{(2)}, l_4^{(2)}]$ for the latches of $F^{(2)}$.

This example demonstrates that the possibility to consider latches as inputs as well as outputs improves our chances to obtain a unique correspondence between the latches for a possible latch permutation equivalence of two sequential circuits. Of course, it is not guaranteed that such a unique possible correspondence will be obtained. It strongly depends on the input and output signatures that we use and on the characteristics of the latches. Thus we focus on the following questions in the rest of this chapter. What input and output signatures can we use? What special properties of the problem do we need to take care of?

2.2 Input signatures

Only some modifications need to be made when considering signatures for input variables as introduced in Chapter 8 for use in the latch correspondence problem.

Let us consider $F \in \mathcal{F}_{n,m,k}$ again. It has m primary outputs, $[y_1, \ldots, y_m]$. We assumed that we can uniquely identify these primary outputs. The corresponding Boolean functions $\lambda_1, \ldots, \lambda_m$ depend on the primary inputs $x_1, \ldots, x_n$ of the sequential circuit and on the additional inputs $u_1, \ldots, u_k$ of the combinational part of the sequential circuit which are due to the removal of the latches. So we can use them step by step, starting with y_1 up to y_m, to compute input signatures for $u_1, \ldots, u_k$.

Let us describe this on an example. Consider F with two input variables, x_1 and x_2, two outputs, y_1 and y_2, and four latches, l_1, l_2, l_3, and l_4. Then, the two outputs y_1 and y_2 represent Boolean functions, namely λ_1 and λ_2, that depend on the set of input variables $[x_1, x_2, u_1, u_2, u_3, u_4]$. Now, let us use the cofactor satisfy count signature function in order to try to separate u_1, u_2, u_3, and u_4. In a first step, we compute this signature using y_1. Note that we must not use one of the outputs v_1, v_2, v_3, or v_4 instead of y_1, since these outputs are not uniquely identified as yet. Let us assume we obtain the partition $[\{u_1, u_3\}, u_4, u_2]$ by applying the cofactor satisfy count signature function with respect to y_1. This means that we can uniquely identify the input variables u_2 and u_4 and thus the latches l_2 and l_4. Furthermore, we obtain a unique partial order of the latches: $[\{l_1, l_3\}, l_4, l_2]$. Now, we can use output function y_2 in a second step to try to distinguish between the latch input variables u_1 and u_3 as well.

2.3 Output signatures

Similar to the input signatures, we can use output signatures to identify the outputs $v_1, \ldots, v_k$ of the combinational part of the sequential circuit. The identification of these outputs implies the identification of the latches as well. Such a *latch output signature function* assigns a value, a vector of values, or a function, which has to be an element of an ordered set, to each output v_j. It provides special information on the corresponding latch l_j considered as output of the combinational part of the sequential circuit. This information has to be independent of any permutation of the latches of F, in general. (Remember that a latch considered as an output of the combinational block usually depends not only on the primary inputs of the sequential circuit, but also on the latches considered as inputs of the combinational block.) Actually, a latch output signature has to be independent of the permutation of those latch input variables only *which have not been uniquely identified yet.* If a latch is uniquely identified during signature computation, this information can be used in the identification process of the remaining latches. This fact is a new aspect compared to conventional permutation independent Boolean comparison as described in Chapter 8. We will go into details in Section 2.3.2.

Now let us develop latch output signatures. For this, let us consider the example circuit $F \in \mathcal{F}_{n,m,k}$ with two input variables, x_1 and x_2, two outputs, y_1 and y_2, and four latches, l_1, l_2, l_3, and l_4, again. Suppose, we could not distinguish between the latches l_1 and l_3 considering them as input variables u_1 and u_3. Thus, we still have the partial order of the latches, $[\{l_1, l_3\}, l_4, l_2]$. Now, let us consider these latches as outputs of the combinational block of F, v_1 and v_3. In general, these outputs represent the transition functions δ_1 and δ_3 that depend on all primary input variables and on the latches considered as input variables.

Now, we can apply the following kinds of latch output signatures to δ_1 and δ_3.

2.3.1 Simple output signatures

All the output signature functions introduced in Chapter 8 Section 3.2 can directly be applied as latch output signatures. The most important ones are

- the satisfy count $|\delta_i|$,

- the vector of the cofactor satisfy count signatures of the input variables of δ_i sorted by the size of the satisfy counts,

$$sort\left(|(\delta_i)_{x_1=1}|, \ldots, |(\delta_i)_{x_n=1}|, |(\delta_i)_{u_1=1}|, \ldots, |(\delta_i)_{u_k=1}|\right),$$

- the breakup signature with respect to function δ_i and origin $o = [0, 0, \ldots, 0]$,

$$[|(\delta_i)^0|, |(\delta_i)^1|, \ldots, |(\delta_i)^{n+k}|].$$

These signature functions satisfy all necessary properties since they are output signatures for δ_i (i.e., for v_i) which are independent of the permutation of the latch input variables, u_1, u_2, u_3, and u_4 as well.

If these simple output signatures do not break the tie, we have to apply some stronger latch output signatures.

2.3.2 Function signatures for latch outputs

Here we use the fact that we have already uniquely identified the primary input variables of the sequential circuit and that some of the latches might have been identified already as well. So, any subfunction of the transition function δ_i that is independent of the latch input variables which are not identified already is a latch output signature. There are several possibilities for this kind of subfunction. Let us consider our example F with the two primary input variables, x_1 and x_2, and the four latch variables considered as inputs, u_1, u_2, u_3, and u_4 again. Subfunctions of the primary output functions δ_1 and δ_3 which only depend on x_1 and x_2 are, for example,

$$(\delta_i)_{[u_1,u_2,u_3,u_4]=(1,1,1,1)},$$

$$(\delta_i)_{[u_1,u_2,u_3,u_4]=(0,0,0,0)},$$

$$(\forall u_4 : (\forall u_3 : (\forall u_2 : (\forall u_1 : \delta_i)))),$$

and

$$(\exists u_4 : (\exists u_3 : (\exists u_2 : (\exists u_1 : \delta_i))))$$

for $i = 1$ and $i = 3$. These function output signatures contain special information about δ_i which is independent of the permutation of the latch input variables of F.

However, as already indicated above, we can choose subfunctions of δ_i, which also depend on latch input variables of latches already identified. We can use exactly the same idea as for constructing the function signatures for input variables (see Chapter 8 on page 181). For each latch which is uniquely identified at this point, the corresponding latch input variable can be uniquely identified as well. In our example, this is the case for the latches l_2 and l_4. So, subfunctions of δ_1 and δ_3 which depend on the primary inputs x_1 and x_2 *and* on the latch input variables u_2 and u_4 are latch output signatures — with one minor restriction: we need to reorder the latch input variables u_2 and u_4 independently of permutation in these subfunctions. Therefore, let us use the order of the latches established by previously used signatures. In the case of our example, the permutation which reorders the input variables of such a subfunction would be (x_1, x_2, u_4, u_2) (see the latch order of our example in the previous section). Such a reordered subfunction of δ_1 and δ_3 has all the properties to be a latch output signature and we can again try to distinguish between δ_1 and δ_3. This process can be iterated as long as we can uniquely identify at least one more latch. That is why we call this extended function signature the *iterating function signature*. Experiments have shown that this is a very powerful signature.

2.3.3 Canonical order signature

There is another strong latch output signature. We call it the *canonical order signature*. For that we use the methods introduced to handle the combinational permutation equivalence problem. Remember that these methods can be used to look for a canonical and permutation independent ordering of the input variables of a Boolean function, up to partial and hierarchical symmetries.

Again, let us consider the Boolean function δ_i which is represented by the output v_i of the combinational block of F. On the input variables of this function, the methods used in Chapter 8 can be applied in order to look for a canonical (up to partial and hierarchical symmetries) permutation independent variable ordering as well. Now, assume that such a canonical permutation independent order exists and suppose that $\pi \in \mathcal{P}er(\mathbb{N}_{n+k})$ is a permutation of the latch input variables of f which constructs this canonical order. Then the canonical order signature is the function

$$\delta_i^{can} = \delta_i \circ \pi.$$

This function is independent of the permutation of the input variables $u_1, \ldots, u_k$ of δ_i. Thus it is a latch output signature. Note that finding this canonical ordering can be restricted to the latch input variables because the primary input variables $x_1, \ldots, x_n$ of δ_i are uniquely identified by assumption.

2.4 An example

In this section, we illustrate our solution paradigm with benchmark *ex4.slif* from the LGSynth91 benchmark set [89]. We are not going to use all signature functions here. However, the general paradigm of finding a unique permutation independent order of the latch variables, as described in the previous sections, will become clear.

Benchmark *ex4.slif* is the description of a sequential circuit with six primary input variables, $x_0, x_1, \ldots, x_5$, nine primary outputs, $y20.14, y20.15, \ldots, y20.22$, and 14 latches. Let us call these latches $l_1, l_2, \ldots, l_{14}$ in the order of their appearance in the benchmark description. For more details please see the benchmark description in the LGSynth91 benchmark set [89].

We begin with computing a *simple output signature*. Let us apply the satisfy count $|v_i|$ of the latch variable l_i considered as output v_i of the combinational block of the sequential circuit. Here, we obtain the following characterization of the latches:

$$\begin{array}{lllll} |v_1| = 0, & |v_2| = 288, & |v_3| = 128, & |v_4| = 128, & |v_5| = 192, \\ |v_6| = 64, & |v_7| = 128, & |v_8| = 128, & |v_9| = 96, & |v_{10}| = 128, \\ |v_{11}| = 128, & |v_{12}| = 192, & |v_{13}| = 64, & |v_{14}| = 128. & \end{array}$$

By these signatures we obtain a partial, permutation independent order of the latches, namely

$$[l_1,\ \underbrace{\{l_6, l_{13}\}}_{\text{group 1}},\ l_9,\ \underbrace{\{l_3, l_4, l_7, l_8, l_{10}, l_{11}, l_{14}\}}_{\text{group 2}},\ \underbrace{\{l_5, l_{12}\}}_{\text{group 3}},\ l_2].$$

That is, there are three aliasing groups of latches, one of size 7 and two of size 2, for which we have to do further computations. We now consider these latches as input variables, u_i, of the combinational block and use an *input signature* to try to distinguish between them. For doing that, we take one primary output function, $y20.i$, after the other and compute the selected input signature with respect to this output function. Let us take the *cofactor satisfy count signature* as input signature and start with primary output function $y20.14$. In this first iteration, we obtain the following results for the three aliasing groups:

aliasing group 1

$$\begin{array}{l} |(y20.14)_{u_6=1}| = 256 \\ |(y20.14)_{u_{13}=1}| = 256, \end{array}$$

aliasing group 2

$$|(y20.14)_{u_3=1}| = 0$$
$$|(y20.14)_{u_4=1}| = 0$$
$$|(y20.14)_{u_7=1}| = 0$$
$$|(y20.14)_{u_8=1}| = 256$$
$$|(y20.14)_{u_{10}=1}| = 0$$
$$|(y20.14)_{u_{11}=1}| = 0$$
$$|(y20.14)_{u_{14}=1}| = 256,$$

and

aliasing group 3

$$|(y20.14)_{u_5=1}| = 128$$
$$|(y20.14)_{u_{12}=1}| = 128.$$

We see that the latches of aliasing group 1 and group 3 cannot be distinguished using the cofactor satisfy count signature with respect to primary output function $y20.14$. However, we can split up group 2 in the subgroups $\{l_3, l_4, l_7, l_{10}, l_{11}\}$ and $\{l_8, l_{14}\}$. So, we obtain a finer partial order for the latches,

$$[l_1,\ \{l_6, l_{13}\},\ l_9,\ \{l_3, l_4, l_7, l_{10}, l_{11}\},\ \{l_8, l_{14}\},\ \{l_5, l_{12}\},\ l_2].$$

At this point we have four aliasing groups of latch variables, namely three of size 2 and one of size 5. Now, we can continue with computing the cofactor satisfy count signatures with respect to primary output function $y20.15$, analyzing the new situation with respect to those signatures (Is there a finer partition of the latches?), and so on for all primary outputs — until there is a unique order of the latches. However, even after using all the primary output variables, we still have just a partial order of the latches,

$$[l_1,\ \{l_6, l_{13}\},\ l_9,\ \{l_4, l_{11}\},\ l_7,\ l_3,\ l_{10},\ l_8,\ l_{14},\ \{l_5, l_{12}\},\ l_2].$$

There are three aliasing groups of size 2. So, let us try and see how the more complex output signatures work. At first, we will use a vector of *function signatures*. This works as follows. We consider the latch variables l_i of the three aliasing groups as outputs v_i of the combinational block of circuit *ex4.slif*. Then we compute subfunctions of the transition function δ_i which is represented by such an output v_i, which are independent of the latch input variables. For the purpose of this example, let us take the following pair of two functions:

$$\left((\delta_i)_{[u_1,\ldots,u_{14}]=(1,\ldots,1)}, (\delta_i)_{[u_1,\ldots,u_{14}]=(0,\ldots,0)}\right).$$

Unfortunately, when applying this signature there is no difference between the function signatures of l_6 and l_{13}, l_4 and l_{11}, and l_5 and l_{12}, respectively.

However, applying the *canonical order signature* helps to distinguish between latch l_4 and l_{11}, and we obtain the partial order

$$[l_1,\ \{l_6, l_{13}\},\ l_9,\ l_{11},\ l_4,\ l_7,\ l_3,\ l_{10},\ l_8,\ l_{14},\ \{l_5, l_{12}\},\ l_2]$$

for the latches.

Now, let us try the *iterating function signature* next. Here, we use the same functions as for the function signature described above, but with one important difference. The computed subfunctions of an output function δ_i are not independent of *all* latch input variables, but only of those which are still in aliasing groups. A vector of such subfunctions of latch variable l_i of one of our aliasing groups is, for example,

$$\left((\delta_i)_{[u_5,u_6,u_{12},u_{13}]=(1,1,1,1)}, (\delta_i)_{[u_5,u_6,u_{12},u_{13}]=(0,0,0,0)}\right).$$

We need to reorder the uniquely identified latch input variables in the restricted functions in order to obtain an iterating function signature. The new order is the permutation independent and unique suborder which is given by the established order of our latch variables

$$(u_1,\ u_9,\ u_{11},\ u_4,\ u_7,\ u_3,\ u_{10},\ u_8,\ u_{14},\ u_2).$$

We finally can establish a unique permutation independent order of all latch variables with the help of this output signature:

$$[l_1,\ l_6,\ l_{13},\ l_9,\ l_{11},\ l_4,\ l_7,\ l_3,\ l_{10},\ l_8,\ l_{14},\ l_5,\ l_{12},\ l_2].$$

Friend, you have read enough. If you desire still more, then be the poem yourself, and all it stands for. — Angelus Silesius

References

[1] M. Abramovici, M. Breuer, and A. Friedman. *Digital Systems Testing and Testable Design Digital Systems Testing and Testable Design (Revised Printing).* IEEE Press, 1990.

[2] A. Aho and J. Ullman. *Foundations of Computer Science*. Computer Science Press, 1992.

[3] S. Akers. Binary decision diagrams. *IEEE Transactions on Computers*, C-27:509–518, 1978.

[4] B. Becker, R. Drechsler, and R. Enders. On the computational power of bit-level and word-level decision diagrams. In *Asia and South Pacific Design Automation Conference*, pages 461–467, 1997.

[5] B. Becker, R. Drechsler, and M. Theobald. On the expressive power of of OKFDDs. *Formal Methods in System Design*, 11(1):5–21, 1997.

[6] B. Becker, R. Drechsler, and R. Werchner. On the relation between BDDs and FDDs. *Information and Computation*, 123(2):185–197, 1995.

[7] L. Benini and G. de Micheli. A survey of boolean matching techniques for library binding. *ACM Transactions on Design Automation and Electronic Systems*, 2:193–226, 1997.

[8] B. Bentley. Validating the Intel Pentium 4 Microprocessor. In *ACM/IEEE Design Automation Conference*, pages 244–249, June 2001.

[9] C. Berman and L. Trevillyan. Functional comparison of logic designs for VLSI circuits. In *International Conference on Computer Aided Design*, pages 456–459, 1989.

[10] A. Biere, A. Cimatti, E. Clarke, M. Fujita, and Y. Zhu. Symbolic model checking using SAT procedures instead of BDDs. In *ACM/IEEE Design Automation Conference*, pages 300–305, 1999.

[11] B. Bollig, P. Savicky, and I. Wegener. On the improvement of variable orderings for OBDDs. In G. Saucier and A. Mignotte, editors, *IFIP Workshop on Logic and Architecture Synthesis*, pages 71–80, 1994.

[12] B. Bollig and I. Wegener. Improving the variable ordering of OBDDs is NP-complete. *IEEE Transactions on Computers*, 45(9):993–1002, 1996.

[13] G. Boole. *The Mathematical Analysis of Logic*. London: G. Bell, 1847.

[14] G. Boole. *An Investigation of the Laws of Thought*. Walton London, 1854. (Reprinted by Dover Books, New York, 1954).

[15] K.S. Brace, R.L. Rudell, and R.E. Bryant. Efficient implementation of a BDD package. In *ACM/IEEE Design Automation Conference*, pages 40–45, June 1990.

[16] D. Brand. Verification of large synthesized designs. In *International Conference on Computer Aided Design*, pages 534–537, 1993.

[17] W. Brauer. *Automatentheorie*. B.G. Teubner Stuttgart, 1984.

[18] R. Brayton, G. Hachtel, C. McMullen, and A. Sangiovanni-Vincentelli. *Logic minimization algorithms for VLSI synthesis*. Kluwer Academic Publishers, 1992.

[19] F. Brown. *Boolean Reasoning - The Logic of Boolean Equations*. Kluwer Academic Publishers, 1990.

[20] R. Bryant. Graph-based algorithms for boolean function manipulation. *IEEE Transactions on Computers*, 35(8):677–691, 1986.

[21] R. Bryant. Binary decision diagrams and beyond: Enabling technologies for formal verification. In *International Conference on Computer Aided Design*, pages 236–243, 1995.

[22] R. Bryant and Y.A. Chen. Verification of arithmetic functions with binary moment diagrams. Technical Report Technical Report CMU-CS-94-160, Carnegie Mellon University, May 1994.

[23] R. Bryant and Y.A. Chen. Verification of arithmetic functions with binary moment diagrams. In *Design Automation Conference*, pages 535–541, 1995.

[24] J. Burch and V. Singhal. Robust latch mapping for combinational equivalence checking. In *International Conference in Computer Aided Design*, pages 563–569, 1998.

[25] J. Burch and V. Singhal. Tight integration of combinational verification methods. In *International Conference on Computer Aided Design*, pages 570–576, 1998.

[26] K. Butler, D. Ross, R. Kapur, and M. Mercer. Heuristics to compute variable oredreings for efficient manipulation of ordered binary decision diagrams. In *Design Automation Conference*, pages 417–420, 1991.

[27] D. I. Cheng and M. Marek Sadowska. Verifying equivalence of functions with unknown input correspondence. In *European Design Automation Conference*, pages 81–85, February 1993.

[28] H. Chockler, O. Kupferman, R.P. Kurshan, and M.Y Vardi. A practical approach to coverage in model checking. In G. Berry, H. Comon, and A. Finkel, editors, *Computer Aided Verification (CAV)*. LNCS 2102, Springer-Verlag Berlin Heidelberg, 2001.

[29] E. Clarke, M. Fujita, and X. Zhao. Hybrid decision diagrams - overcoming the limitations of MTBDDs and BMDs. In *International Conference of Computer Aided Design*, 1995.

[30] E.M. Clarke and E.A. Emerson. Characterizing properties of parallel programs as fixpoints. In *Seventh International Colloquium on Automata, Languages, and Programming*, 1981.

[31] T. Cormen, C. Leiserson, and R. Rivest. *Introduction to algorithms*. The MIT Press, 1990.

[32] O. Coudert and J. Madre. A unified frameqork for the formal verification of sequential circuits. In *International Conference on Computer-Aided Design*, pages 126–129, 1990.

[33] W. Daehn. *Testverfahren in der Mikroelektronik*. Springer Verlag, 1997.

[34] M. Davis, G. Logemann, and D. Loveland. A machine program for theorem-proving a machine program for theorem-proving. *Communications of the ACM*, 5(7):394–397, July 1962.

[35] M. Davis and H. Putnam. A computing procedure for quantification theory. *Journal of the Association for Computer Machinery (JACM)*, 7:201–215, 1960.

[36] G. de Micheli. *Synthesis and Optimization of Digital Circuits*. McGraw-Hill New Yorkj, 1994.

[37] R. Dechter and I. Rish. Directional resolution: The davis-putnam procedure, revisited. In *KR - Principles of Knowledge Representation and Reasoning*, 1994.

[38] D. Dempster and M. Stuart. *Verification methodology manual - Techniques for verifying HDL designs*. Teamwork International, 2001.

[39] M.L. Dertouzos. *Threshold Logic: A Synthesis Approach*. MIT Cambridge, 1965.

[40] R. Drechsler. *Ordered Kronecker Functional Decision Diagrams und ihre Anwendung*. PhD thesis, Department of Computer Science, Wolfgang Goethe-Universitätt Frankfurt am Main, Germany, 1996. (In German).

[41] R. Drechsler. Fast exact minimization of BDDs. In *Design Automation Conference*, pages 200–205, 1998.

[42] R. Drechsler. *Formal Verification of Circuits*. Kluwer Academic Publishers Boston, Dordrecht, London, 2000.

[43] R. Drechsler and B. Becker. *Binary Decision Diagrams – Theory and Implementation*. Kluwer Academic Publishers Boston, Dordrecht, London, 1998.

[44] R. Drechsler and W. Günther. Using lower bounds during dynamic BDD minimization. In *Design Automation Conference*, pages 29–32, 1999.

[45] R. Drechsler and S. Höreth. Manipulation of *BMDs. In *Asia and South Pacific Design Automation Conference*, pages 433–438, 1998.

[46] R. Drechsler, A. Sarabi, M. Theobald, B. Becker, and M. Perkowski. Efficient representation and manipulation of switching functions based on ordered kronecker functional decision diagrams. In *Design Automation Conference*, pages 415–419, 1994.

[47] R. Drechsler and D. Sieling. Binray decision diagrams in theory and practice. *International Journal on Software Tools for Technology Transfer*, 2(3):112–136, 2001.

[48] E. Emerson. *Handbook of Theoretical Computer Science: Formal Methods and Semantics*, chapter Temporal and modal logic, pages 995–1072. MIT Press, 1990.

[49] J.W. Freeman. *Improvements to Propositional Satisfiability Search Algorithms*. PhD thesis, University of Pennsylvania, 1995.

[50] S. Friedman and K. Supowit. Finding the optimal variable ordering for binary decision diagrams. *IEEE Transactions on Computer-Aided Design*, 39(5):710–713, 1990.

[51] H. Fujii, G. Ootomo, and C. Hori. Interleaving based variable ordering methods for ordered binary decision diagrams. In *International Conference on Computer Aided Design*, pages 38–41, 1993.

[52] M. Fujita, Y. Matsunaga, and T. Kakuda. On variable ordering of binary decision diagrams for the application of multi-level logic synthesis. In *European Conference on Design Automation*, pages 50–54, 1991.

[53] M. Garey and D. Johnson. *Computers and Intractability - A Guide to the Theory of NP-Completeness*. W.H. Freeman and Company, 1979.

[54] W. Günther, N. Drechsler, R. Drechsler, and B. Becker. Verification of designs containing black boxes. In *Proceediings of EUROMICRO*, pages 100–105, 2000.

[55] A. Gupta and P. Ashar. Integrating a boolean satisfiability checker and BDDs for combinational equivalence checking. In *International Conference ov VLSI Design*, pages 222–225, 1997.

[56] G.D. Hachtel and F. Somenzi. *Logic synthesis and verification algorithms*. Kluwer Academic Publishers Boston, Dordrecht, London, 1996.

[57] K. Hamaguchi, A. Morita, and S. Yajima. Efficient construction of binary moment diagrams for verifying arithmetic circuits. In *International Conference on Computer-Aided Design*, pages 78–82, 1995.

[58] H. Hengster. *EXOR-basierte Logiksynthese mit Kronecker Functional Deciion Diagrams*. PhD thesis, Institute for Computer Science, Albert-Ludwigs-University Freiburg, Germany, 2000.

[59] M. Herbstritt and B. Becker. Conflict-based selection of branching rules in SAT-algorithms. In R. Drechsler, editor, *Methoden und Beschreibungssprachen zur Modellierung und Verifikation von Schaltungen und Systemen*, pages 189–198. Shaker Verlag Aachen, February 2003.

[60] A. Hett. *Binäre Expression-Diagramme*. PhD thesis, Institute for Computer Science, Albert-Ludwigs-University Freiburg, Germany, 2002.

[61] J. Hooker and V. Vinay. Branching rules for satisfiability. GSIA Working Paper 1994-09, Graduate School of Industrial Administration, Carnegie-Mellon University, Pittsburgh, Pennsylvania, July 1994.

[62] J. Hopcroft, R. Motwani, and J. Ullman. *Introduction to Automata Theory, Languages, and Computation*. Addison-Wesley, 2001.

[63] G. Hotz. *Schaltungstheorie*. De Gruyter Lehrbuch, Walter De Gruyter, 1974.

[64] H. Hulgaard, P. Williams, and H. Andersen. Equivalence checking of combinational circuits using boolean expression diagrams. *IEEE Transactions on Computer-Aided Design*, 18(7), 1999.

[65] S.L. Hurst, D.M. Miller, and J.C. Muzio. *Spectral Techniques in Digital Logic*. Academic London, 1985.

[66] N. Ishiura, H. Sawada, and S. Yajima. Minimization of binary decision diagrams based on exchanges of variables. In *International Conference on Computer Aided Design*, pages 472–475, 1991.

[67] A. Jain, V. Boppana, R. Mukherjee, J. Jain, M. Fujita, and M. Hsiao. Testing, verification, and diagnosis in the presence of unknowns. In *Proceedings of VLSI Test Symposium*, pages 263–269, 2000.

[68] S. Jeong, T. Kim, and F. Somenzi. An efficient method for optimal BDD ordering computation. In *International Conference on VLSI and CAD*, 1993.

[69] R. Jeroslow and J. Wang. Solving propositional satisfiability problems. *Annals of Mathematics and Artificial Intelligence*, pages 167–187, 1990.

[70] R. Katz. *Contemporary Logic Design*. The Benjamin/Cummings Publishing Company, 1994.

[71] S. Katz, D. Geist, and O. Grumberg. 'Have I written enough properties?' - a method of comparison between specification and implementation. In L. Pierre and T. Kropf, editors, *CHARME*, pages 280–297. LNCS 1703, Springer-Verlag Berlin Heidelberg, 1999.

[72] U. Kebschull and W. Rosenstiel. Efficient graph-based computation and manipulation of functional decision diagrams. In *European Conference on Design Automation*, pages 278–282, 1993.

[73] U. Kebschull, E. Schubert, and W. Rosenstiel. Multilevel logic synthesis based on functional decision diagrams. In *European Conference on Design Automation*, pages 43–47, 1992.

[74] M. Keim, M. Martin, P. Molitor, B. Becker, and R. Drechsler. Polynomial formal verification of multipliers. *Formal Methods in System Design*, 22:39–58, 2003.

[75] Z. Kohavi. *Switching and Finite Automata Theory*. McGraw-Hill, 1970.

[76] R. Kolla and P. Molitor. Logic synthesis of digital circuits. Series of lectures at University Würzburg and University Halle, 2000.

[77] A. Kuehlmann and F. Krohm. Equivalence checking using cuts and heaps. In *Design Automation Conference*, pages 263–268, 1997.

[78] A. Kuehlmann, V. Paruthi, F. Krohm, and M. Ganai. Robust boolean reasoning for equivalence checking and functional property verification. *IEEE Transactions on Computer-Aided Design*, 21(12):1377–1393, 2002.

[79] A. Kuehlmann and C. van Eijk. *Logic Synthesis and Verification* (Editors: S. Hassoun and T. Sasao), chapter Combinational and Sequential Equivalence Checking, pages 343–371. Kluwer Academic Publishers, 2002.

[80] W. Kunz, J. Marques-Silva, and S. Malik. *Logic Synthesis and Verification* (Editors: S. Hassoun and T. Sasao), chapter SAT and ATPG: Algorithms for Boolean Decision Problems, pages 309–341. Kluwer Academic Publishers, 2002.

[81] W. Kunz and D. Pradhan. Recursive learning: A new implication technique for efficient solutions to CAD-problems. *IEEE Transactions on Computer-Aided Design*, 13(9):1143–1158, 1994.

[82] W. Kunz and D. Stoffel. *Reasoning in Boolean Networks*. Kluwer Academic Publishers, 1997.

[83] W. Kunz and D. Stoffel. Equivalence checking using structural methods. *it+ti - Informationstechnik und Technische Informatik*, 1, February 2001. In German.

[84] Y. Lai and S. Sastry. Edge-valued binary decision diagrams for multi-level hierarchical verification. In *Design Automation Conference*, pages 608–613, 1992.

[85] Y.-T. Lai, S. Sastry, and M. Pedram. Boolean matching using binary decision diagrams with applications to logic synthesis and verification. In *International Conference on Computer Design*, pages 452–458, October 1992.

[86] T. Larrabee. Test pattern generation using boolean satisfiability. *IEEE Transactions on Computer-Aided Design*, 11(1):4–15, January 1992.

[87] R.J. Lechner. *Harmonic Analysis of Switching Functions*, chapter Recent Developments in Switching Theory. Academic New York, 1971.

[88] C. Lee. Representation of switching circuits by binary decision diagrams. *Bell Syst. Tech. Journal*, 38:985–999, 1959.

[89] LGSynth91. Combinational and sequential logic benchmark suite. International Workshop on Logic Synthesis (IWLS), 1991.

[90] C. Li and X. Anbulagan. Heuristics based on unit propagation for satisfiability. In *International Joint Conference on Artificial Intelligence (IJCAI)*, pages 315–326, 1997.

[91] F. Mailhot and G. de Micheli. Technology mapping using boolean matching and don't care sets. In *European Design Automation Conference*, pages 212–216, February 1990.

[92] F. Mailhot and G. de Micheli. Algorithms for technology mapping based on binary decision diagrams and on boolean operations. *IEEE Transactions on Computer-Aided Design*, 12(5):599–620, 1993.

[93] S. Malik, A. R. Wang, R. K. Brayton, and A. Sangiovanni-Vincentelli. Logic verification using binary decision diagrams in a logic synthesis environment. In *International Conference on Computer Aided Design*, pages 6–9, November 1988.

[94] J. Marques-Silva. The impact of branching heursitics in propositional satisfiability algorithms. In *9th Portuguese Conference on Artificial Intelligence (EPIA)*, September 1999.

[95] J. Marques-Silva and T. Glass. Combinational equivalence checking using satisfiability and recursive learning. In *Design, Automation, and Test in Europe*, pages 145–149, 1999.

[96] J. Marques-Silva and K. Sakallah. Conflict analysis in search algorithms or propositional satisfiability. In *IEEE International Conference on Tools with Artificial Intelligence*, 1996.

[97] J. Marques-Silva and K. Sakallah. GRASP: A search algorithm for propositional satisfiability. *IEEE Transactions on Computers*, 48(5):506–521, May 1999.

[98] Y. Matsunaga. An efficient equivalence checker for combinational circuits. In *33rd Design Automation Conference*, pages 629–634, 1996.

[99] E. McCluskey. *Logic Design Principles*. Prentice Halle, 1986.

[100] K.L. McMillan. *Symbolic Model Checking*. Kluwer Academic Publishers Boston, Dordrecht, London, 1993.

[101] C. Meinel. Binäre Entscheidungsgraphen. Lecture Notes, Department of Computer Science, University Trier, Germany, 1997.

[102] C. Meinel and A. Slobodova. Speeding up variable reordering of OBDDs. In *International Conference on Computer Design*, pages 338–343, 1997.

[103] M. Mercer, R. Kapur, and D. Ross. Functional approaches to generating orderings for efficient symbolic representations. In *Design Automation Conference*, pages 624–627, 1992.

[104] S. Minato, N. Ishiura, and S. Yajima. Shared binary decision diagram with attributed edges for efficient boolean function manipulation. In *Design Automation Conference*, pages 52–57, 1990.

[105] J. Mohnke. *A Signature-based Approach to Formal Logic Verification*. PhD thesis, Institute for Computer Science, Martin-Luther-University Halle, D-06120 Halle (Saale), Germany, 1999.

[106] J. Mohnke and S. Malik. Permutation and phase independent boolean comparison. *INTEGRATION - the VLSI journal*, 16:109–129, 1993.

[107] J. Mohnke, P. Molitor, and S. Malik. Establishing latch correspondence for sequential circuits using distinguishing signatures. *INTEGRATION, the VLSI Journal*, 27:33–46, 1999.

[108] J. Mohnke, P. Molitor, and S. Malik. Application of BDDs in boolean matching techniques for formal logic combinational verification. *International Journal on Software Tools for Technology Transfer*, 2(3):207–216, 2001.

[109] J. Mohnke, P. Molitor, and S. Malik. Limits of using signatures for permutation independent boolean comparison. *Formal Methods in System Design*, 21(2):98–112, September 2002. An extended abstract has been presented at the ASP-DAC 1995.

[110] D. Möller, J. Mohnke, and M. Weber. Detection of symmetry of boolean functions represented by ROBDDs. In *International Conference on Computer Aided Design*, pages 680–684, 1993.

[111] B. Moret. Decision trees and diagrams. *Comput. Surveys*, 14:593–623, 1982.

[112] M. Moskewicz, C. Madigan, Y. Zhao, L. Zhang, and S. Malik. Chaff: Engineering an efficient SAT solver. In *Design Automation Conference (DAC)*, pages 530–535, 2001.

[113] F. Ozguner, D. Marhefka, J. DeGroat, B. Wile, J. Stofer, and L. Hanrahan. Teaching future verification engineers: The forgotten side of logic design. In *Design Automation Conference*, 2001.

[114] S. Panda and F. Somenzi. Who are the variables in your neighborhood ? In *International Conference on Computer Aided Design*, pages 74–77, 1995.

[115] S. Panda, F. Somenzi, and V. Plessier. Symmetry detection and dynamic variable ordering of decision diagrams. In *International Conference on Computer Aided Design*, pages 628–631, 1994.

[116] D. Patterson and J. Hennessy. *Computer Organization & Design – The Hardware / Software Interface*. Morgan Kaufmann Publishers, 1998.

[117] D. Paul, M. Chatterjee, and D. Pradhan. Verilat: Verification using logic augmentation and transformations. *IEEE Transactions on Computer Aided Design*, 19(9):1041–1051, September 2000.

[118] M. Payer. Industrial experience with formal verification. *it+ti Informationstechnik und Technische Informatik, Oldenbourg Verlag München*, 2001(1):16–21, 2001.

[119] I. Pomeranz and S.M. Reddy. On diagnosis and correction of design errors. In *International Conference oc Computer Aided Design*, pages 500–507, November 1993.

[120] I. Pomeranz and S.M. Reddy. On determining symmetries in inputs of logic circuits. *IEEE Transactions on Computer-Aided Design*, 13(11):1428–1434, November 1994.

[121] S. Quer, G. Cabodi, P. Camurati, L. Lavagno, E.M. Sentovich, and R.K. Brayton. Incremental fsm re-encoding for simplifying verification by symbolic traversal. In *International Workshop on Logic Synthesis*, May 1995.

[122] R. Ranjan, A. Aziz, R. Brayton, B. Plessier, and C. Pixley. Efficient BDD algorithms for FSM synthesis and verification. In *IEEE/ACM International Workshop on Logic Synthesis*, 1995.

[123] S. Reda and A. Salem. Combinational equivalence checking using boolean satisfiability and binary decision diagrams. In *Design, Automation, and Test in Europe*, pages 122–126, 2001.

[124] J. Roth. Diagnosis of automata failures: A calculus and a method. *IBM Journal of Research and Development*, 10(4):278–291, July 1966.

[125] R. Rudell. Dynamic variable ordering for ordered binary decision diagrams. In *International Conference on Computer Aided Design*, pages 42–47, 1993.

[126] T. Sasao and M. Fujita. *Representations of Discrete Functions*. Kluwer Academic Publishers, 1996.

[127] U. Schlichtmann, F. Brglez, and P. Schneider. Efficient boolean matching based on unique variable ordering. In *International Workshop on Logic Synthesis*, May 1993.

[128] C. Scholl. *Functional Decomposition with Application to FPGA Synthesis*. Kluwer Academic Publishers, Boston / Dordrecht / London, 2001.

[129] C. Scholl and B. Becker. Checking equivalence for partial implementations. Technical Report 145, Institute of Computer Science at the Albert-Ludwigs-University Freiburg, October 2000.

[130] C. Scholl and B. Becker. Checking equivalence for partial implementations. In *38th Design Automation Conference*, pages 238–243, 2001.

[131] C. Scholl, B. Becker, and T.M. Weis. Word-level decision diagrams, WLCDs and division. In *International Conference on Computer Aided Design*, pages 672–677, 1998.

[132] C. Scholl, D. Möller, P. Molitor, and R. Drechsler. BDD minimization using symmetries. *IEEE Transactions on Computer-Aided Design*, 18(2):81–100, 1999.

[133] M. Schulz and E. Auth. Improved determinsitic test pattern generation with applications to redundancy identification. *IEEE Transactions on Computer-Aided Design*, 8(7):811–816, July 1989.

[134] M. Schulz, E. Trischler, and T. Sarfert. SOCRATES: A highly efficient automatic test pattern generation system. In *International Test Conference (ITC)*, pages 1016–1026, 1987.

[135] M. Schulz, E. Trischler, and T. Sarfert. SOCRATES: A highly efficient automatic test pattern generation system. *IEEE Transactions on Computer-Aided Design*, 7(1):126–137, January 1988.

[136] P. Tafertshofer and A. Ganz. SAT based ATPG using fast justification and propagation in the implication graph. In *International Conference on Computer Aided Design*, pages 139–146, 1999.

[137] P. Tafertshofer, A. Ganz, and M. Henftling. A SAT-based implication engine for efficient ATPG, equivalence checking, and optimization of netlists. In *International Conference on Computer Aided Design*, pages 648–655, 1997.

[138] T. Thierauf. *The Computational Complexity of Equivalence and Isomorphism Problems*. PhD thesis, University Ulm, Germany, 1998.

[139] T. Thierauf. The isomorphism problem for read-once branching programs and arithmetic circuits. *Chicago Journal on Theoretical Computer Science*, 1998.

[140] C. van Eijk. *Formal Methods for the Verification of Digital Circuits Formal Methods for the Verification of Digital Circuits*. PhD thesis, Eindhoven University of Technology, 1997.

[141] J. van Eijk. Sequential equivalence checking without state space traversal. In *Design, Automation, and Test in Europe*, pages 618–623, 1998.

[142] C. Wallace. A suggestion for a fast multiplier. *IEEE*, 13:14–17, 1964.

[143] K. Wang, T. Hwang, and C. Chen. Restructering binary decision diagrams based on functional equivalence. In *European Design Automation Conference*, pages 261–265, February 1993.

[144] K. Wang, T. Hwang, and C. Chen. Exploiting communication complexity for boolean matching. *IEEE Transactions on Computer-Aided Design*, 15(10):1249–1256, 1996.

[145] S. Wefel and P. Molitor. Prove that a faulty multiplier is faulty !? In *Great Lakes Symposium on VLSI*, 2000.

[146] I. Wegener. *The Complexity of Boolean Functions*. John Wiley & Sons Ltd., and B.G. Teubner, Stuttgart (Wiley–Teubner Series in Computer Science), 1987.

[147] I. Wegener. *Branching programs and binary decision diagrams – theory and applications*. SIAM Discrete Mathematics and Applications, 2000.

[148] H. Zhang. SATO: An efficient propositional prover. In *International Conference on Automated Deduction*, pages 272–275, July 1997.

[149] L. Zhang, C. Madigan, M. Moskewicz, and S. Malik. Efficient conflict driven learning in a boolean satisfiability solver. In *International Conference on Computer Aided Design (ICCAD)*, pages 279–285, 2001.

Index

About the Authors

Prof. Dr. Paul Molitor (*1959) studied computer science at the University of Saarland, Germany, from 1978 to 1986. He received a Diploma, a Dr.rer.nat., and a Dr.habil degree in 1982, 1986, and 1992, respectively. From 1982 to 1992 he was with the research group on VLSI design methods and parallelism at the University of Saarland. After serving as a Visiting Professor at the University of Halle in 1992/93 and the University of Freiburg in 1993, he was an Associate Professor in the Computer Science Department of the Humboldt University Berlin in 1993/94. Since 1994, he has been a Full Professor at the Martin-Luther-University Halle-Wittenberg, Germany. Dr. Molitor is author of a German textbook on Data Structures and Efficient Algorithms for Logic Synthesis intended for graduate students. He published about 50 international papers in the field of modelling, physical design, logic synthesis, and formal verification of digital circuits and in the field of the development of hardware algorithms. Currently, he serves as Editor-in-Chief of the oldest German scientific journal on information technology, *it - Information Technology*.

Dr. Janett Mohnke (*1967) studied computer science at Humboldt University in Berlin, Germany from 1986 to 1991. She received a Diploma in 1991. From 1991 to 1994 she was with the research group on VLSI design methods at Humboldt University. During this time she spent six month at the Electrical Engineering Department of Princeton University in Princeton, USA. From 1995 to 1997 she joined the VLSI research group at Martin-Luther-University in Halle-Wittenberg, Germany. Since 1997 she has been with DResearch Digital Media Systems. Janett Mohnke received her PhD in 1999. The topic of her thesis includes permutation independent Boolean comparison and the latch correspondence problem.

About the Authors